Fractional Calculus in Medical and Health Science

Mathematics and its Applications
Modelling, Engineering, and Social Sciences
*Series Editor: Hemen Dutta, Department of Mathematics,
Gauhati University*

Tensor Calculus and Applications
Simplified Tools and Techniques
Bhaben Kalita

Discrete Mathematical Structures
A Succinct Foundation
Beri Venkatachalapathy Senthil Kumar and Hemen Dutta

Methods of Mathematical Modelling
Fractional Differential Equations
Edited by Harendra Singh, Devendra Kumar, and Dumitru Baleanu

Mathematical Methods in Engineering and Applied Sciences
Edited by Hemen Dutta

Sequence Spaces
Topics in Modern Summability Theory
Mohammad Mursaleen and Feyzi Başar

Fractional Calculus in Medical and Health Science
Edited by Devendra Kumar and Jagdev Singh

For more information about this series, please visit: www.crcpress.com/Mathematics-and-its-applications/book-series/MES

ISSN (online): 2689-0224
ISSN (print): 2689-0232

Fractional Calculus in Medical and Health Science

Edited by
Devendra Kumar
Jagdev Singh

CRC Press is an imprint of the
Taylor & Francis Group, an **informa** business

MATLAB® is a trademark of The MathWorks, Inc. and is used with permission. The MathWorks does not warrant the accuracy of the text or exercises in this book. This book's use or discussion of MATLAB® software or related products does not constitute endorsement or sponsorship by The MathWorks of a particular pedagogical approach or particular use of the MATLAB® software.

First edition published 2020
by CRC Press
6000 Broken Sound Parkway NW, Suite 300, Boca Raton, FL 33487-2742

and by CRC Press
2 Park Square, Milton Park, Abingdon, Oxon, OX14 4RN

© 2021 Taylor & Francis Group, LLC

CRC Press is an imprint of Taylor & Francis Group, LLC

Reasonable efforts have been made to publish reliable data and information, but the author and publisher cannot assume responsibility for the validity of all materials or the consequences of their use. The authors and publishers have attempted to trace the copyright holders of all material reproduced in this publication and apologize to copyright holders if permission to publish in this form has not been obtained. If any copyright material has not been acknowledged please write and let us know so we may rectify in any future reprint.

Except as permitted under U.S. Copyright Law, no part of this book may be reprinted, reproduced, transmitted, or utilized in any form by any electronic, mechanical, or other means, now known or hereafter invented, including photocopying, microfilming, and recording, or in any information storage or retrieval system, without written permission from the publishers.

For permission to photocopy or use material electronically from this work, access www.copyright.com or contact the Copyright Clearance Center, Inc. (CCC), 222 Rosewood Drive, Danvers, MA 01923, 978-750-8400. For works that are not available on CCC please contact mpkbookspermissions@tandf.co.uk

Trademark notice: Product or corporate names may be trademarks or registered trademarks, and are used only for identification and explanation without intent to infringe.

Library of Congress Cataloging-in-Publication Data

Names: Kumar, Devendra, editor.
Title: Fractional calculus in medical and health science / edited by Devendra Kumar and Jagdev Singh.
Description: Boca Raton, FL : CRC Press/Taylor & Francis Group, [2021] | Series: Mathematics and its applications | Includes bibliographical references and index.
Identifiers: LCCN 2020007783 (print) | LCCN 2020007784 (ebook) | ISBN 9780367351212 (hardback ; acid-free paper) | ISBN 9780429340567 (ebook)
Subjects: LCSH: Medicine--Mathematics. | Fractional calculus.
Classification: LCC R853.M3 F69 2021 (print) | LCC R853.M3 (ebook) | DDC 610.1/5195--dc23
LC record available at https://lccn.loc.gov/2020007783
LC ebook record available at https://lccn.loc.gov/2020007784

ISBN: 978-0-367-35121-2 (hbk)
ISBN: 978-0-429-34056-7 (ebk)

Typeset in Palatino
by Lumina Datamatics Limited

Contents

Preface vii
Editors xi
Contributors xiii

1. **Image Edge Detection Using Fractional Conformable Derivatives in Liouville-Caputo Sense for Medical Image Processing** 1
   J. E. Lavín-Delgado, J. E. Solís-Pérez, J. F. Gómez-Aguilar and R. F. Escobar-Jiménez

2. **WHO Child Growth Standards Modelling by Variable-, Fractional-Order Difference Equation** 55
   Piotr Ostalczyk

3. **Fractional Calculus Approach in SIRS-SI Model for Malaria Disease with Mittag-Leffler Law** 83
   Jagdev Singh, Sunil Dutt Purohit and Devendra Kumar

4. **Mathematical Modelling and Analysis of Fractional Epidemic Models Using Derivative with Exponential Kernel** 109
   Kolade M. Owolabi and Abdon Atangana

5. **Fractional Order Mathematical Model for the Cell Cycle of a Tumour Cell** 129
   Ritu Agarwal, Kritika and Sunil Dutt Purohit

6. **Fractional Order Model of Transmission Dynamics of HIV/AIDS with Effect of Weak CD4+ T Cells** 149
   Ved Prakash Dubey, Rajnesh Kumar and Devendra Kumar

7. **Fractional Dynamics of HIV-AIDS and Cryptosporidiosis with Lognormal Distribution** 167
   M. A. Khan and Abdon Atangana

v

8. **A Fractional Mathematical Model to Study the Effect of Buffer and Endoplasmic Reticulum on Cytosolic Calcium Concentration in Nerve Cells** 211
   Brajesh Kumar Jha and Hardik Joshi

9. **Fractional SIR Epidemic Model of Childhood Disease with Mittag-Leffler Memory** 229
   P. Veeresha, D. G. Prakasha and Devendra Kumar

Index 249

Preface

This book is extensively developed for graduate students, medical practitioners, and researchers working in the area of fractional calculus and epidemiological modelling of infectious diseases. Fractional calculus and mathematical modelling find many real life applications in biology, medical science, physics, chemistry economics, mechanical engineering, computer science, signal processing, etc. This book presents several applications of fractional calculus in medical and health science. It consists of nine chapters and is organised as follows.

Chapter 1 presents a novel fractional conformable edge detector for medical images structure feature extraction. The case studies are detection and analysis of cerebral arteriovenous malformations, meningiomas, and medulloblastomas by using computerised tomograph scans (CTs) and magnetic resonance imaging scans (MRIs), which is motivated by the fact that these medical images are most commonly sought out for clinical study to diagnose different diseases or pathologies for establishing treatment planning. In this sense, for a given medical image, the aim of the this chapter is to take advantage of the frequency characteristics of the fractional conformable derivative for extracting more structure and feature details of such images, thus achieving, to give patients a timely diagnosis and adequate medical treatment. The experimental evaluation has demonstrated that the proposed operator gives superior performance compared to those classic operators, Gabor method and a fuzzy operator because it is able to detect more edge details and features of the medical images. Also, the proposed operator is more robust to noise. For the first time, in this chapter, novel fractional conformable masks based on fractional conformable derivative with power-law for images edge detection are presented.

In Chapter 2, the discrete linear time – invariant variable-, fractional-order model of living organisms population growth represented by percentile charts containing the rates: weight vs. age. Fractional models described by the difference equations containing fractional order Grünwald-Letnikow backward differences of commensurate or non-commensurate orders show better matching to the measured data. In the chapter, a generalisation of the mentioned difference: the variable- and fractional-order backward difference is applied. In place of the fractional order, there is applied a discrete function of order. This leads to further improvement of the mathematical model. Some numerical examples supporting the considerations are also given.

Chapter 3 contains a new investigation of fractional SIRS-SI model of malaria disease and it examines the transmission of malaria in human body. Malaria is a serious life-threatening disease. As per the data of WHO, in the year 2017, in among 87 countries, there were approximately 219 million malaria cases that reportedly arose. According to WHO, approximately, a number of 435,000 people have died due to malaria in the year 2017. In this chapter, the authors discuss SIRS-SI malaria model by employing Atangana and Baleanu operator having strong memory effect. The authors have discussed about the treatment of malaria and to control it, some suggestions are also given. To analyse the existence and uniqueness of the solution of SIRS-SI model with non-singular and non-local kernel, the authors have applied the fixed-point theorem. It is shown that the treatment for malaria has made a huge impact on both mosquito and human population. The numerical results of SIRS-SI malaria disease model with non-integer order are obtained by applying new powerful technique namely q-HATM. The Maple package is used to demonstrate the impact of various parameters on the treatment of malarial disease. The numerical outcomes show that for the arbitrary order SIRS-SI model of malaria disease and the discussed method is very strong, effective and computationally very easy.

Chapter 4 examines the two important epidemic models with delay terms involving Caputo-Fabrizio operator. In the models, the classical time derivatives are replaced with their fractional counterparts. The system is first examined for local and global stability analysis to ensure that correct choice of parameters is chosen during numerical simulations. By employing the fixed-point theorem, the conditions for the existence and uniqueness of solutions in the models are established. The effect of fractional parameter is verified and displayed in figures to show the behaviour of the species.

Chapter 5 presents a mathematical model for tumour growth with the help of biology behind the cell cycle. This model provides the dynamics of population of three tumour cells such as quiescent cells, cells midst interphase, and mitotic cells. This chapter studies the fractional extension of the model by using the Caputo-Fabrizio derivative of arbitrary order possessing non-singular kernel. The iterative perturbation technique is used to solve this tumour cell development model. The fixed-point theorem has been used to show the existence and uniqueness of the solution. Some numerical solution has been provided and illustrated graphically to view the impact of fractionalisation of the model.

Chapter 6 deals with the application of Adams-Bashforth-Moulton (ABM) method to the fractional order model of transmission dynamics of HIV/AIDS with the effect of weak CD4+ T cells. The generalised ABM method, usually interpreted as the fractional Adams method, is an advantageous numerical scheme to handle the fractional ordinary differential equations. The numerical solution is derived through the proposed numerical method to analyse the variations of uninfected cells, infected cells and virus cells with respect to time and varying values of fractional order of time derivative. The fractional derivative in this model is considered in Caputo sense.

Chapter 7 provides a detailed analysis of the dynamical process of the spread of one of the more fatal infectious disease known as HIV-AIDS with the inclusion of cryptosporidiosis using different approaches. On one hand, the authors have investigated the model by using different concepts of differentiation and integration. On the other hand, the authors investigated a model within the framework of stochastic analysis by using an unusual method called log-normal distribution. The chapter provides a motivation for using each concept, for instance, the authors argued that the spread can follow at the same time a Gaussian and non-Gaussian distribution but with a steady state, which corresponds to the probability distribution associated to the exponential decay law used to construct the well-known Caputo-Fabrizio fractional differential operator. However, if the distribution does not have a steady state, from Gauss to non-Gauss, the distribution is linked to that of the generalised Mittag-Leffler function used in the Atangana-Baleanu fractional derivative, thus the model with non-singular and non-local kernel. Finally if the model displays some random behaviours but yet follows Markovian process, then the stochastic approach is adopted. For each model, the authors have provided a detailed analysis underpinning the determination of conditions under which the existence of unique set of exact solutions are guaranteed and finally provides numerical solutions by using a newly established numerical scheme similar to Adams-Bashforth, which is yet more efficient and user friendly. Theoretical and numerical simulations suggest that these mathematical models could be used to better understand the dynamics behind the spread of HIV/AIDS.

Chapter 8 presents a one-dimensional fractional diffusion equation to study the effect of buffers and endoplasmic reticulum on calcium distribution profile in nerve cells. Various physiological parameters like association rate, diffusion coefficients, buffering rate constant, endoplasmic reticulum flux are incorporated in the model to study their effect on the intracellular calcium concentration. Fractional Laplace and Fourier transforms are employed to solve the mathematical model. The obtained results are simulated in MATLAB® and interpreted with neuronal disorders like Parkinson's disease. It is observed that the buffer has a significant effect on cytosolic calcium profile, whereas endoplasmic reticulum plays an important role in maintaining the free calcium ion profile.

Chapter 9 presents a detailed analysis of the susceptible-infected-recovered epidemic model of childhood disease with Atanagana-Baleanu fractional operator. The considered non-linear model has been efficiently applied to describe the evolution of childhood disease in a population and its influence on the community. The Adams-Bashforth scheme is applied to find and analyse the solution for the proposed model. The fixed-point hypothesis considered in order to demonstrate the existence and uniqueness of the derived solution for the future fractional order model. Two distinct explanatory cases are considered and for both cases, the simulations have been demonstrated in terms of plots. The present investigation shows that the Atanagana-Baleanu derivative plays a vital role in the analysis and description of the behaviour of diverse models arising in the study of human diseases.

The results presented in this book will be very useful for mathematicians, scientists, doctors, and engineers working on fractional calculus and its applications in mathematical modelling of medical and health science problems.

Devendra Kumar
Jaipur, India

Jagdev Singh
Jaipur, India

MATLAB® is a registered trademark of The MathWorks, Inc. For product information, please contact:

The MathWorks, Inc.
3 Apple Hill Drive
Natick, MA 01760-2098 USA
Tel: 508 647 7000
Fax: 508-647-7001
E-mail: info@mathworks.com
Web: www.mathworks.com

Editors

Devendra Kumar is an assistant professor in the department of mathematics, University of Rajasthan, Jaipur, Rajasthan, India. He did his master of science (MSc) in mathematics and PhD in mathematics from the University of Rajasthan, India. He primarily teaches subjects such as real and complex analysis, functional analysis, integral equations, and special functions in a postgraduate-level course in mathematics. His areas of interest are mathematical modelling, special functions, fractional calculus, applied functional analysis, non-linear dynamics, analytical, and numerical methods. He has published three books. His works have been published in the *Nonlinear Dynamics, Chaos* Solitons *& Fractals, Physica A, Journal of Computational and Nonlinear Dynamics, Applied Mathematical Modelling, Entropy, Advances in Nonlinear Analysis, Romanian Reports in Physics, Applied Mathematics and Computation, Chaos* and several other peer-reviewed international journals. He has published 191 research papers in various reputed journals with an h-index of 38. He has attended a number of national and international conferences and presented several research papers. He has also attended summer courses, short-term programs, and workshops. He is a member of the editorial board and reviewer of various mathematical journals.

Jagdev Singh is a professor in the department of mathematics, JECRC University, Jaipur, Rajasthan, India. He did his master of science (MSc) in mathematics and PhD in mathematics from the University of Rajasthan, India. He primarily teaches subjects such as mathematical modelling, real analysis, functional analysis, integral equations, and special functions in a postgraduate-level course in mathematics. His areas of interest are mathematical modelling, mathematical biology, fluid dynamics, special functions, fractional calculus, applied functional analysis, non-linear dynamics, analytical, and numerical methods. He has published four books. His works have been published in the *Nonlinear Dynamics, Chaos Solitons & Fractals, Physica A, Journal of Computational and Nonlinear Dynamics, Applied Mathematical Modelling, Entropy, Advances in Nonlinear Analysis, Romanian Reports in Physics, Applied Mathematics and Computation, Chaos* and several other peer-reviewed international journals. He has published 175 research papers in various reputed journals with an h-index of 39. He has attended a number of national and international conferences and presented several research papers. He has also attended summer courses, short-term programs, and workshops. He is a member of the editorial board and a reviewer of various mathematical journals.

Contributors

Ritu Agarwal
Department of Mathematics
Malaviya National Institute of Technology
Jaipur, India

Abdon Atangana
Institute for Groundwater Studies
Faculty of Natural and Agricultural Sciences
University of the Free State
Bloemfontein, South Africa

Ved Prakash Dubey
Faculty of Mathematical and Statistical Sciences
Shri Ramswaroop Memorial University
Lucknow, India

R. F. Escobar-Jiménez
Tecnológico Nacional de México/CENIDET
Interior Internado Palmira S/N
Cuernavaca, México

J. F. Gómez-Aguilar
CONACyT-Tecnológico Nacional de México/CENIDET
Interior Internado Palmira S/N
Cuernavaca, México

Brajesh Kumar Jha
Department of Mathematics
School of Technology
Pandit Deendayal Petroleum University
Gandhinagar, India

Hardik Joshi
Department of Mathematics
School of Technology
Pandit Deendayal Petroleum University
Gandhinagar, India

M. A. Khan
Department of Mathematics
City University of Science and Information Technology
Peshawar, Pakistan

and

Institute for Groundwater Studies
Faculty of Natural and Agricultural Sciences
University of the Free State
Bloemfontein, South Africa

Kritika
Department of Mathematics
Malaviya National Institute of Technology
Jaipur, India

Devendra Kumar
Department of Mathematics
University of Rajasthan
Jaipur, India

Rajnesh Kumar
Department of Applied Science and Humanities
Government Engineering College
Nawada, India

and

Department of Science and Technology
Bihar, India

J. E. Lavín-Delgado
Tecnológico Nacional de México/CENIDET
Interior Internado Palmira S/N
Cuernavaca, México

Piotr Ostalczyk
Institute of Applied Computer Science
Lodz University of Technology
Łódź, Poland

Kolade M. Owolabi
Faculty of Mathematics and Statistics
Ton Duc Thang University
Ho Chi Minh City, Vietnam

D. G. Prakasha
Department of Mathematics
Davangere University
Davangere, India

Sunil Dutt Purohit
Department of HEAS (Mathematics)
Rajasthan Technical University
Kota, India

Jagdev Singh
Department of Mathematics
JECRC University
Jaipur, India

J. E. Solís-Pérez
Tecnológico Nacional de México/CENIDET
Interior Internado Palmira S/N
Cuernavaca, México

P. Veeresha
Department of Mathematics
Karnatak University
Dharwad, India

1

Image Edge Detection Using Fractional Conformable Derivatives in Liouville-Caputo Sense for Medical Image Processing

J. E. Lavín-Delgado, J. E. Solís-Pérez, J. F. Gómez-Aguilar and R. F. Escobar-Jiménez

CONTENTS

1.1 Introduction 1
1.2 Mathematical Preliminaries 5
1.3 Edge Detection Using Fractional Conformable Derivative 6
    1.3.1 Fractional Conformable Gaussian Kernel 7
    1.3.2 Fractional Conformable Gaussian Gradient 8
1.4 Performance Test..9
1.5 Test Images...21
    1.5.1 Cerebral Arteriovenous Malformation.......................................21
    1.5.2 Meningioma...26
    1.5.3 Medulloblastoma 30
    1.5.4 Abdominal Aortic Aneurysm 32
    1.5.5 Cerebral Venous Infarction 35
    1.5.6 Breast Calcifications...39
1.6 Discussion and Conclusion 48
Acknowledgements 49
Competing Interests...49
Authors' Contributions 49
References..50

1.1 Introduction

In computer vision and image processing, edge detection refers to the process of identifying and locating points in a digital image at which the image brightness changes sharply or, more formally, has discontinuities. These points define the boundaries between regions in an image, which are very useful for many applications, such as 3D reconstruction, object recognition, motion analysis, pattern recognition, medical image processing, image enhancement and restoration, and so on. Edge detection simplifies the amount of data to be processed and filters out useless information while preserving the important structural properties of the image. Medical image processing and analysis has received considerable attention recently because they are arduous to process due to their complexity and their distinct modalities [1]. Therefore, researchers and the medical practitioners, in many cases, cannot accurately detect and diagnose the diseases in conventional ways [2]. In this way, there should be a system that helps them to understand medical images very easily. Image segmentation using edge detection can be used for analysis and better visualisation of medical images, so that the processed images can easily be performed using the analysis the image data, thereby enhancing the ability to study, monitor, diagnose, and treat disorders or diseases. In this context, several challenges have been proposed within the area of medical image processing and analysis (including the edge detection) through group research and medical associations [3]. All of them focussed on improving the detection, diagnosis, and treatment of diseases. In the literature, several edge detectors have been proposed, which differ in their mathematical and algorithmic properties, but most of them can be classified into two main categories: search based and zero-crossing based [4]. The search-based methods detect edges by computing the first-order derivatives to find the changes of intensity, such as Roberts [5], Prewitt [6], Sobel [7], Canny [8] and Chen [9]. The zero-crossing-based methods search for zero crossings in a second-order derivative expression computed from the image in order to find edges, such as presented in Marr and Hildreth [10] and Haralick [11]. As a pre-processing step to edge detection, a smoothing stage is almost always applied, usually by using the Gaussian filter. In recent years, there has been considerable interest in fractional calculus of several fields of science, including physics, chemistry, biology, signal processing, robotics, and control theory because it generalises mathematical models described by differential equations, obtaining more accurate representations than the integer-order methods [12]. So, it is easy to imagine why many fractional-order-based methods have been used for image enhancement, image denoising, image restoration, edge and corner detection, etc. [13]. In [14], the discrete fractional Hilbert transform was developed and applied for edge detection in digital images. The numerical simulations showed that the proposed discrete fractional Hilbert transform can be successfully used to detect edges and corners of digital images. The work presented in [15] showed how introducing an edge detector based on non-integer (fractional) differentiation can improve the criterion of 'thin detection' or detection selectivity in the case of parabolic luminance transitions, and the criterion of immunity to noise, which can be interpreted in terms of robustness to noise in general. An edge detection operator that considers a fractional differentiation and integration approach was proposed in [16]. The fractional operator had a good performance in terms of detection accuracy and noise robustness. In [17], a fractional differential mask for edge detection was proposed. The fractional mask was based on the Riemann-Liouville derivative. The tests showed that this mask can not only maintain the low-frequency contours information in the smooth regions but also enhance the high-frequency edges and texture part in the image. This property generates a visual effect for the images whose texture information has an important meaning. In tests carried out in 1D samples, the fractional mask showed good performance in terms of detection effectiveness and noise immunity, whereas that in tests carried out in 2D samples, the results indicated that without image noise, the proposed mask can accurately detect edges; meanwhile, with image noise it can effectively suppress noises [18]. A fractional-order gradient operator for medical image structure feature extraction is described in [19]. The method can be seen as generalisation of the first-order Sobel operator based on the Grünwald-Letnikov derivative definition. The results showed that the proposed fractional-order operator yields good visual effects. In [20], an edge detector based on fractional-order differentiation was introduced, which can significantly improve the detection performance to noisy images. Furthermore, this operator is implemented in a modular railway track measurement system. In [21], a 1D digital fractional-order Charef differentiator was introduced and extended to 2D-fractional differentiation for edge detection. The mask coefficients were computed in a way that image details are detected and preserved. The obtained results over texture images demonstrated the efficiency of the proposed operator compared to classic techniques. In [22], an edge detector operator was introduced, which is based on Gaussian kernel smoothing, fractional partial derivatives, and a statistical approach. The model was validated on different types of textured images from the Brodatz album. The work described in [23] proposes a novel method for edge detection based on the Grünwald-Letnikov fractional-order derivative and the fractional-order Fourier transformation, that is, this edge detection operator incorporates the fractional-order differentiation in the fractional Fourier transform domain. Through experimental simulations it is shown that this technique can detect the edges precisely and efficiently. In addition, their performance is better than other existing methods in terms of robustness to noise. In [24], a fractional-order edge detector operator with contrast enhancement of images based on Riemann-Liouville fractional derivative is proposed. By theoretical and experimental results, it is observed that the proposed operator outperforms the traditional methods through well-known metrics such as Pratt Figure of Merit. These results also showed that the fractional-order derivative improves the texture and contrast of the image. A contour segmentation method to segment blood vessels images based on fractional-order differentiation and fuzzy energy is developed in [22]. First, as for the blurry boundaries, a fractional-order differential method is used to enhance the original image for accurate segmentation. Then, to deal with intensity inhomogeneity, it incorporates the fuzzy local statistical information into the model by the usage of a Gaussian kernel function for contours segmentation. Experimental results demonstrate a desirable performance of the proposed method for blood vessel images in terms of accuracy and robustness. In [25], an optimisation technique to select the best threshold levels to enhance the edge detection algorithms based on fractional mask is proposed. From the peak signal-to-noise ratio (PSNR) and Bit Error Rate (BER) results, the performance comparison shows that the algorithms using the optimisation technique based on fractional edge detection are better than the fractional edge detection algorithms, as they get the optimal threshold levels for different types of images. The work presented in [26] develops a fractional-order differential mask for edge detection. The texture information of the test images is enhanced, so that complete and continuous edges are obtained with this proposed mask. In addition, the smooth texture information is also retained. An edge detection algorithm of remote-sensing images based on the fractional-order Chebyshev polynomial is presented in [27]. The algorithm has been validated with the benchmark data set provided by International Society for Photogrammetry and Remote Sensing set working group. Experimental results showed that the proposed algorithm is better with respect to traditional edge-detection techniques. These results also showed that a higher order of differentiator gives more accurate edge detection. In [28], a fractional-order edge detection operator for the detection of Alzheimer's disease from an MRI scan of the brain is proposed. The mean square error (MSE) and PSNR are used for performance comparison with classic methods. Finally, this fractional-order operator can be an aid for the early detection of Alzheimer's disease. A modified Grünwald-Letnikov derivative to enhance more and detect better the edges of an image is presented in [29]. Experimental simulations showed that the proposed operator can be efficiently employed in different areas of image processing, such as image enhancement, edge detection, and medical diagnostic. In [24], a fractional differential operator for feature and contrast enhancement of images using the Riemann-Liouville derivative was implemented. The results with over six input images showed that the proposed method outperforms classical edge-detection operators for texture and contrast enhancement. In the previously mentioned works, the authors considered different fractional-order derivatives whose frequency responses preserve low-frequency contour features in smooth regions, while at the same time keep high-frequency features where gray level changes frequently, and also enhance medium-frequency texture details [30].

In the preceding literature, the authors considered the Riemann-Liouville-Caputo fractional derivative with power-law singular kernel [31–34]. Recently, some researchers introduced the concept of a non-local derivative. In [35], Khalil presented the 'conformable derivative'. This definition is compatible with the classical derivative and satisfied some conventional properties, for instance, the chain rule. Based on this definition and involving the Riemann-Liouville-Caputo fractional derivative, the authors in [36] introduced new fractional integration and differentiation operators. The operators obtained in the Riemann-Liouville-Caputo and Hadamard sense were obtained iterating conformable integrals. Furthermore, authors proved that these fractional integrals and derivatives have similar properties to the ordinary Newton derivative. Because these operators depend on two fractional parameters (the conformable and the fractional), we obtain better detection of the memory. In our paper, the first parameter (conformable) can be related with the fractal contour. This characteristic allows the operator to describe systems with different spatial scales, and the second parameter (fractional) is related to the texture of the images.

The main contribution of the present research is the developing of a fractional conformable-order edge detector iterating conformable integrals. Using visual perception and statistical analysis, the proposed novel fractional conformable mask with two orders shows significant advantages over other fractional edge detectors presented in the literature. The developing of fractional masks involving the fractional conformable derivative with two orders has not been reported in the literature yet. The choice of addressing this issue is due to the great interest of the scientific community in the image-processing analysis.

1.2 Mathematical Preliminaries

Definition 1.1

The Liouville-Caputo operator (C) is the convolution of the local derivative of a given function with power-law function. The Liouville-Caputo fractional derivative of order ($\alpha > 0$) is defined as follows [31]:

$$ {}_a^C \mathcal{D}_t^\alpha f(t) = \frac{1}{\Gamma(n-\alpha)} \int_a^t \frac{d^n}{d\theta^n} f(\theta)(t-\theta)^{n-\alpha-1} d\theta, \qquad n-1 < \alpha < n. \qquad (1.1) $$

Definition 1.2

According to [35], it should be $f : (0, \infty) \to \Re$, then, the conformable derivative of $f(t)$ with order ($\alpha > 0$) is given by:

$$ {}_a^K \mathcal{D}_t^\alpha f(t) = \lim_{\varepsilon \to 0} \frac{f(t + \varepsilon t^{1-\alpha}) - f(t)}{\varepsilon}, \qquad (1.2) $$

for all $t > 0$, $\alpha \in (0, 1)$. If $f(t)$ is α-differentiable in some $(0, a)$, $a > 0$, $\lim_{t \to 0^+} f^\alpha(t)$ and exists, then we define $f^\alpha(0) = \lim_{t \to 0^+} f^\alpha(t)$

Definition 1.3

Let $\operatorname{Re}(\beta) \geq 0$, $n = \left[\operatorname{Re}(\beta)\right] + 1$, $f \in C_{a,t}^n([a,t])$. Then, the fractional conformable derivative in the Liouville-Caputo sense is given by [36]

$$\begin{aligned}{}_a^c\mathcal{D}_t^\beta{}_t^\alpha f(t) &= \frac{1}{\Gamma(n-\beta)} \int_a^t \left(\frac{(t-a)^\alpha - (x-a)^\alpha}{\alpha}\right)^{n-\beta-1} \frac{{}_a^n\mathcal{D}_t^\alpha f(x)}{(x-a)^{1-\alpha}} dx, \\ &= {}^{cn-\beta}_a I_t^\alpha\left({}^{cn}_a\mathcal{D}_t^\alpha f(t)\right).\end{aligned} \quad (1.3)$$

Definition 1.4

The classical Gaussian distribution in 1D is expressed as

$$g(x) = \frac{1}{\sigma\sqrt{2\pi}} \exp\left(-\frac{(x-\mu)^2}{2\sigma^2}\right), \quad (1.4)$$

where μ is the mean, $\sigma > 0$ is the standard deviation, and x is the input. In this proposal, μ is centred on $\mu = 0$, that is, it has a zero mean.

1.3 Edge Detection Using Fractional Conformable Derivative

The fractional derivatives have similar properties to Newton's derivative, and they have been applied successfully to extract hidden information from some complex systems. On the other hand, in spite of the conformable derivatives not being a fractional derivative, formally speaking, their non-integer order satisfies several properties from the classical calculus. Both features are present in the fractional conformable derivatives because of the two fractional orders, α and β. These parameters allow not only provide two freedom degrees but also get a better detection of the memory. According to [37], the memory means that the current dynamic or behaviour is determined by past dynamics with special forms of weights. In Figure 1.1, the numerical simulation of the power-law function and Khalil-type conformable operator involved in fractional conformable derivative for different values of α, β arbitrarily chosen is shown. In these behaviours, one can see that the exponential decay is fast or slow dependent on the values of α and β chosen.

Image Edge Detection Using Fractional Conformable Derivatives

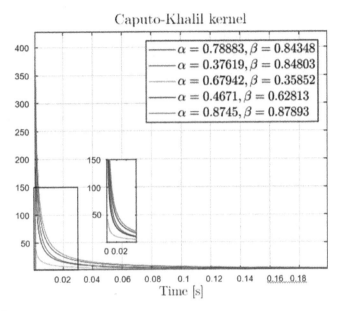

FIGURE 1.1
Fractional conformable kernel for different α, β values arbitrarily chosen.

1.3.1 Fractional Conformable Gaussian Kernel

In order to get the fractional conformable Gaussian kernel, first we have to derive the classical Gaussian kernel in Equation (1.4). Then, the classical Gaussian kernel derivative is given as follows:

$$\frac{dg(x)}{dx} = -\frac{(x-\mu)}{\sigma^3\sqrt{2\pi}}\exp\left(-\frac{(x-\mu)^2}{2\sigma^2}\right). \tag{1.5}$$

By generalising the classical operator d/dx in Equation (1.5) with the fractional conformable operator of Khalil type in Caputo sense in Equation (1.3), the following expressions are obtained:

- Fractional conformable Gaussian kernel along x

$$^{c}_{0}\mathcal{D}^{\alpha}_{t}{}^{\beta}g(x) = t^{1-\alpha}\left[-\frac{(x-\mu)}{\sigma^3\sqrt{2\pi}}\exp\left(-\frac{(x-\mu)^2}{2\sigma^2}\right)\right], \quad \begin{array}{l} 0<\alpha<1, \\ 0<\beta<1. \end{array} \tag{1.6}$$

- Fractional conformable Gaussian kernel along y

$$^{c}_{0}\mathcal{D}^{\alpha}_{t}{}^{\beta}g(y) = t^{1-\alpha}\left[-\frac{(y-\mu)}{\sigma^3\sqrt{2\pi}}\exp\left(-\frac{(y-\mu)^2}{2\sigma^2}\right)\right], \quad \begin{array}{l} 0<\alpha<1, \\ 0<\beta<1. \end{array} \tag{1.7}$$

The numerical solutions to the fractional conformable Gaussian kernels in Equations (1.6) and (1.7) are obtained from the scheme proposed in [38]. This scheme is based on the Adams-Moulton algorithm and the trapezoidal quadrature rule.

Hence, let us consider $h = \frac{T}{N}$, xj, $nh = 0, 1, 2, 3,..., N$ with N steps in an interval of [0, T] to discretise Equations (1.6) and (1.7). The Adams-Moulton method for the fractional conformable Gaussian kernel of the Khalil type in the Caputo sense along x and y are given as follows:

$$g_h^p(x_n+1) = g_{x0} + \frac{1}{\Gamma(\beta)} \sum_{j=0}^{n} b_{j,n+1} \mathcal{D}_*^\alpha g(x_j), \quad \begin{matrix} 0 < \alpha < 1, \\ 0 < \beta < 1, \end{matrix} \quad (1.8)$$

$$g_h^p(y_n+1) = g_{y0} + \frac{1}{\Gamma(\beta)} \sum_{j=0}^{n} b_{j,n+1} \mathcal{D}_*^\alpha g(y_j), \quad \begin{matrix} 0 < \alpha < 1, \\ 0 < \beta < 1, \end{matrix} \quad (1.9)$$

where α denotes the order of the conformable derivative in Equation (1.2), β represents the fractional order of the Caputo fractional derivative and $\mathcal{D}_*^\alpha$ is the conformable derivative proposed by Khalil

$$\mathcal{D}_*^\alpha g(x_j) = (t^{1-\alpha}) \frac{d g(x_j)}{dx}, \quad (1.10)$$

$$\mathcal{D}_*^\alpha g(y_j) = (t^{1-\alpha}) \frac{d g(y_j)}{dy}, \quad (1.11)$$

moreover,

$$b_{j,n+1} = \frac{h^\beta}{\beta} \left[(n+1-j)^\beta - (n-j)^\beta \right], \quad 0 \leq j \leq n. \quad (1.12)$$

1.3.2 Fractional Conformable Gaussian Gradient

If an image is represented as an image function $f(x, y)$, where (x, y) is the spatial coordinate and f is the pixel intensity at (x, y), then it is possible to get two or more directions, such as the horizontal, the vertical or a combination of them defined as an arbitrary direction.

In order to get the fractional conformable Gaussian gradient defined as $\nabla I^{\alpha,\beta}$, we represent it as a vector derivative description

$$\nabla I^{\alpha,\beta} = \frac{\partial^\alpha I}{\partial x} \hat{i}_x + \frac{\partial^\alpha I}{\partial y} \hat{i}_y = \left({}^{KC}g_x^{\alpha,\beta} \otimes I \right) \hat{i}_x + \left({}^{KC}g_x^{\alpha,\beta} \otimes I \right) \hat{i}_y, \quad (1.13)$$

where $\otimes$ denotes the convolution operator, I is the input image, and i_x and i_y are the unit vectors in horizontal and vertical directions, respectively. Also, ${}^{KC}g_x^{\alpha,\beta}$ and ${}^{KC}g_y^{\alpha,\beta}$ denote the horizontal and vertical fractional conformable Gaussian derivative filter given by

$$ {}^{KC}g_x^{\alpha,\beta} = {}_0^c\mathcal{D}_t^\beta\, {}_t^\alpha g(x) \otimes \begin{bmatrix} 1 & 0 & -1 \end{bmatrix}, \tag{1.14}$$

$$ {}^{KC}g_y^{\alpha,\beta} = {}_0^c\mathcal{D}_t^\beta\, {}_t^\alpha g(y) \otimes \begin{bmatrix} 1 \\ 0 \\ -1 \end{bmatrix}. \tag{1.15}$$

where ${}_0^c\mathcal{D}_t^\beta\, {}_t^\alpha g(x)$ and ${}_0^c\mathcal{D}_t^\beta\, {}_t^\alpha g(y)$ depict the fractional conformable Gaussian kernel along x and y given by Equations (1.6) and (1.7), respectively.

The edge detection method based on the fractional conformable filter derivatives in Equations (1.14) and (1.15) is summarised in the next algorithm.

Algorithm 1.1 Edge detection using fractional gradient

1: The input image is converted to gray scale (I)
2: $\alpha, \beta \in (0, 1)$ and $\sigma \in [1, 2.5]$ are chosen to generate the fractional conformable derivative filters
3: The image detected edges are obtained by the following expression:

$$ {}^{KC}_x\mathcal{E}_y^{\alpha,\beta} = \mathrm{abs}\left({}^{KC}g_x^{\alpha,\beta} \otimes I\right) + \mathrm{abs}\left({}^{KC}g_y^{\alpha,\beta} \otimes I\right). \tag{1.16}$$

NOTE: The numerical solution to the Equations (1.14) and (1.15) involved in Equation (1.16) are obtained by Equations (1.8) through (1.12).

Therefore, the proposed edge detector is conformable because it has the Khalil-type operator in Equation (1.2) and at the same time fractional due to the power-law kernel in Equation (1.1).

1.4 Performance Test

The performance assessment of an image edge detector is an important task to do before being implemented. The most commonly used metrics for assessing the quality of images edges are the MSE, the PSNR and structural similarity index (SSIM) [39]. MSE is basically a weighted function of deviations in images, or the square difference between compared images is basically $I_1(x, y)$ and a weighted $I_2(x, y)$ given by [40]

$$\text{MSE} = \frac{1}{mn}\sum_{i=1}^{m}\sum_{j=1}^{n}\left[I_1(i,j) - I_2(i,j)\right]^2, \quad (1.17)$$

where m and n stand for image size. Another measure closely related to MSE and which also indicates the level of losses or signals integrity is the next one [41]

$$\text{PSNR} = 10\log\left[\frac{\max(I)^2}{\text{MSE}}\right]. \quad (1.18)$$

Although these measures are widely used, they are not completely suitable for understanding the actual quality of the edges. In this sense, SSIM is more slightly related to the human visual system, as it extracts useful information, such as luminance l, contrast c, and structure s [42]. It has been used successfully to evaluate structure preservation and the noise removal. SSIM similarity measure is given by

$$\text{SSIM} = \left[l(I_1, I_2)\right]^{\zeta} \cdot \left[c(I_1, I_2)\right]^{\delta} \cdot \left[s(I_1, I_2)\right]^{\gamma}, \quad (1.19)$$

where

$$l(I_1, I_2) = \frac{2\mu I_1 \mu I_2 + C_1}{\mu_{I_1}^2 + \mu_{I_2}^2 + C_1}, \quad c(I_1, I_2) = \frac{2\sigma I_1 \sigma I_2 + C_2}{\sigma_{I_1}^2 + \sigma_{I_2}^2 + C_2},$$

$$s(I_1, I_2) = \frac{\sigma I_1 \sigma I_2 + C_3}{\sigma_{I_1}^2 + \sigma_{I_2}^2 + C_3}.$$

C_1, C_2 and C_3 are small constants that are included to avoid instability when the denominator is very close to zero; μI_1, μI_2 are the means; $\sigma I_1 I_2$ is the covariance; and σI_1, σI_2 are the standard deviations of the ideal image (I_1) and the filtered image (I_2), respectively. In addition, ζ, δ, and γ are positive parameters that control the relative importance of the components. The main limitation of SSIM measure is the inability to measure highly blurred images successfully [43]. Edge-strength-similarity-based image metric (ESSIM) is an improved measure, based on SSIM, which compares the edge information between the analysed images [44]. Let be the ideal image (ground truth) as

$$I_1 = [I_{1_1}, \ldots, I_{1_i}, \ldots, I_{1_N}] \in \mathbb{R}^N,$$

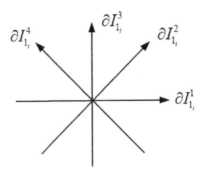

FIGURE 1.2
Directional derivatives.

where N is the total number of pixels. The directional derivative at the ith pixel, ∂I_i^j, indicates the local regularity of the image along the j-direction [43]. For simplicity, only the directions shown in Figure 1.2 are considered, and they are computed by convolving the image with the following kernels Kj based on Scharr operator [44]

$$K^1 = \frac{1}{16}\begin{bmatrix} 0 & 0 & 0 & 0 & 0 \\ 0 & -3 & 0 & 3 & 0 \\ 0 & -10 & 0 & 10 & 0 \\ 0 & -3 & 0 & 3 & 0 \\ 0 & 0 & 0 & 0 & 0 \end{bmatrix}, \quad K^2 = \frac{1}{16}\begin{bmatrix} 0 & 0 & 0 & 3 & 0 \\ 0 & 0 & 0 & 10 & 0 \\ -3 & 0 & 0 & 0 & 3 \\ 0 & -10 & 0 & 0 & 0 \\ 0 & 0 & -3 & 0 & 0 \end{bmatrix},$$

$$K^3 = \frac{1}{16}\begin{bmatrix} 0 & 0 & 0 & 3 & 0 \\ 0 & 3 & 10 & 3 & 0 \\ 0 & 0 & 0 & 0 & 0 \\ 0 & -3 & -10 & -3 & 0 \\ 0 & 0 & -3 & 0 & 0 \end{bmatrix}, \quad K^4 = \frac{1}{16}\begin{bmatrix} 0 & 0 & 3 & 0 & 0 \\ 0 & 10 & 0 & 0 & 0 \\ 3 & 0 & 0 & 0 & -3 \\ 0 & 0 & 0 & -10 & 0 \\ 0 & 0 & -3 & 0 & 0 \end{bmatrix}.$$

Hence, the edge-strength in the diagonal and vertical (or horizontal) directions, respectively, are defined as

$$E_i^{2,4}(I_1) = \left|\partial I_{1_i}^2 - \partial I_{1_i}^4\right|^p,$$

$$E_i^{1,3}(I_1) = \left|\partial I_{1_i}^1 - \partial I_{1_i}^3\right|^p,$$

(1.20)

where p is a non-linear scaling factor that adjusts the edge-strength to the human visual system. Thus, the total edge-strength is given by

$$E(I_1,i) = \max\left[E_i^{1,3}(I_1), E_i^{2,4}(I_1)\right]. \tag{1.21}$$

If the filtered image of I_1 is denoted as

$$I_2 = [I_{2_1}, \ldots, I_{2_i}, \ldots, I_{2_N}] \in \mathbb{R}^N.$$

then the edge-strength of I_2 at the ith pixel is defined as

$$E(I_2,i) = \begin{cases} E_i^{1,3}(I_2) & \text{if } E(I_1,i) = E_i^{1,3}(I_1), \\ E_i^{2,4}(I_2) & \text{if } E(I_1,i) = E_i^{2,4}(I_1), \end{cases}$$

where $E_i^{1,3}(I_2)$ and $E_i^{2,4}(I_2)$ are defined in similar way as in Equation (1.20). The visual fidelity between I_1 and I_2 can be calculated by the similarity between the largest edge-strength, so that the ESSIM is given by

$$\text{ESSIM}(I_1, I_2) = \frac{1}{N} \sum_{i=1}^{N} \frac{2E(I_1,i)E(I_2,i) + C}{\left[E(I_1,i)\right]^2 + \left[E(I_2,i)\right]^2 + C}, \tag{1.22}$$

where the parameter C is a scaling parameter. For the 8-bit greyscale images, this value is calculated as

$$C = (255 B_1)^{2p},$$

where $B_1 = B^{\frac{1}{p}}$. In [44], p is chosen to be 0.5, while B_1 is 10. The performance of the proposed fractional conformable operator against the traditional techniques commonly used for edge detection is evaluated in this section by considering four images taken from the Berkeley Segmentation Dataset and Benchmark [45]. An ideal robust edge detector should generate the same edge map for both noisy and noiseless images, hence the proceeding for calculating MSE and PSNR are considered to select an original image, and then add noise to this image of two different kinds: salt and pepper noise with a density of $d = 0.1$, and Gaussian noise with mean of $m = 0$ and variance $v = 0.01$. Finally, the classical operators and proposed fractional conformable operator are implemented to the noisy images, and Equation (1.17) is applied to calculate the MSE values. Similarly, Equation (1.18) is used to compute the PSNR values. In order to determine the effect of the conformable order α

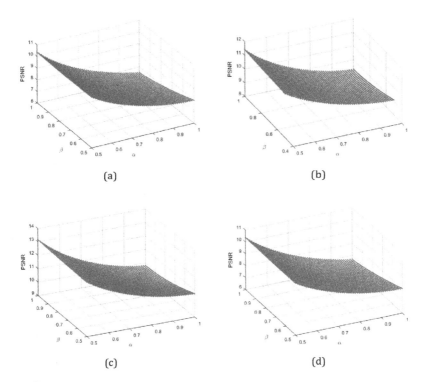

FIGURE 1.3
PSNR analysis: (a) birds, (b) owl, (c) wildcat, and (d) street.

and the fractional order β for edge detection, different values of them were examined. Figure 1.3 shows that high values of PSNR are obtained for small values of α and β; however, too much texture is detected and the strong edges disappear. According to this figure and to maps of edges obtained for the different values of α and β analysed, it is concluded that the conformable order is closely related to the edges generated by the boundaries of objects, while the fractional order affects the weak edges due to the texture of the image. A good choice of these values depends on the type of image, as well as, in the task that the proposed edge detector will be applied. In this work, the values used are $\alpha = 0.94$, $\beta = 0.91$ and $\sigma = 1.45$ due to the proposed operator having a good performance in terms of the quality of the edges detected and robustness to noise (PSNR).

The output edge-detected images obtained and the values of MSE and PSNR are show in Tables 1.1, 1.2 and 1.4–1.7 for $\alpha = 0.94$, $\beta = 0.91$ and $\sigma = 1.45$. According to the obtained results, the fractional conformable operator performs better than any of traditional operators for edge detection in noisy images, that is, it is more robust to noise than other techniques. On the other hand, ground truth edge maps taken from a Berkeley segmentation

TABLE 1.1

Edge Detection Operators Implemented in Images Corrupted with Salt and Pepper Noise

	Birds	Owl	Wildcat	Street
Original image				
Noisy image (salt and pepper)				
Sobel				
Roberts				
LoG				
Prewitt				
Canny				
Gabor				
Fuzzy				
Fractional conformable (1.16) $\alpha = 0.94$ $\beta = 0.91$ $\sigma = 1.45$				

Image Edge Detection Using Fractional Conformable Derivatives 15

TABLE 1.2

Edge Detection Operators Implemented in Images Corrupted with Gaussian Noise

	Birds	Owl	Wildcat	Street
Original image				
Noisy image				
Sobel				
Roberts				
LoG				
Prewitt				
Canny				
Gabor				
Fuzzy				
Fractional conformable (1.16) $\alpha = 0.94$ $\beta = 0.91$ $\sigma = 1.45$				

TABLE 1.3

Comparison between Edge-Detected Images and Ground Truth Edge Maps

	Birds	Owl	Wildcat	Street
Original image				
Ground truth image				
Sobel				
Roberts				
LoG				
Prewitt				
Canny				
Gabor				
Fuzzy				
Fractional conformable (1.16) $\alpha = 0.94$ $\beta = 0.91$ $\sigma = 1.45$				

dataset [45] were considered to develop the ESSIM analysis. The analysis results are shown in Tables 1.3 through 1.7, the numerical analysis shows that the proposed operator detects the most important image structure information. In conclusion, the fractional conformable operator performs better than the classical methods, such as Sobel, Roberts, LoG, Prewitt, and Canny,

TABLE 1.4
Performance Parameters of Birds Image

		Salt and Pepper Noise		Gaussian Noise		
Birds	Operator	MSE	PSNR(dB)	MSE	PSNR(dB)	ESSIM
	Sobel	15805.612670	6.142690	15809.522716	6.141616	0.9953
	Roberts	15808.490210	6.141899	15814.533940	6.140239	0.9953
	LoG	15796.232407	6.145268	15783.204130	6.148851	0.9954
	Prewitt	15805.702573	6.142665	15809.537088	6.141612	0.9953
	Canny	15791.177110	6.146658	15753.832112	6.156941	0.9955
	Gabor	15973.272886	6.096864	16468.557742	5.964247	0.9956
	Fuzzy	15956.477082	6.101433	16451.094118	5.968855	0.9950
	Fractional conformable (1.16) $\alpha = 0.94, \beta = 0.91$ $\sigma = 1.45$	**11867.578228**	**7.387182**	**11517.285943**	**7.517302**	**0.9957**

TABLE 1.5
Performance Parameters of Owl Image

		Salt and Pepper Noise		Gaussian Noise		
Owl	Operator	MSE	PSNR(dB)	MSE	PSNR(dB)	ESSIM
	Sobel	14010.415593	6.666293	14012.341604	6.665696	0.9911
	Roberts	14016.406661	6.664436	14020.014468	6.663318	0.9909
	LoG	13992.506492	6.671848	13991.975518	6.672013	0.9915
	Prewitt	14010.908342	6.666140	14012.578305	6.665623	0.9911
	Canny	13983.549381	6.674629	13982.088153	6.675083	0.9917
	Gabor	14171.859599	6.616535	14507.550456	6.514862	0.9892
	Fuzzy	1440.193962	6.546593	14486.386655	6.521202	0.9905
	Fractional conformable (1.16) $\alpha = 0.94, \beta = 0.91$, $\sigma = 1.45$	**8606.627450**	**8.782473**	**8592.291941**	**8.789713**	**0.9909**

TABLE 1.6

Performance Parameters of Wildcat Image

Wildcat	Operator	Salt and Pepper Noise MSE	PSNR(dB)	Gaussian Noise MSE	PSNR(dB)	ESSIM
	Sobel	8756.595145	8.707450	8758.472328	8.706519	0.9925
	Roberts	8759.331247	8.706094	8762.991865	8.704279	0.9925
	LoG	8743.018270	8.714189	8740.496965	8.715442	0.9928
	Prewitt	8756.736258	8.707380	8758.484446	8.706513	0.9925
	Canny	8734.755299	8.718296	8728.485579	8.721414	0.9930
	Gabor	9004.699153	8.586111	9398.193360	8.400359	0.9931
	Fuzzy	9244.712417	8.471869	9380.183450	8.408690	0.9924
	Fractional conformable (1.16) $\alpha = 0.94$, $\beta = 0.91$ $\sigma = 1.45$	**6110.293910**	**10.270182**	**6002.987377**	**10.347129**	0.9936

TABLE 1.7

Performance Parameters of Street Image

Street	Operator	Salt and Pepper Noise MSE	PSNR(dB)	Gaussian Noise MSE	PSNR(dB)	ESSIM
	Sobel	16182.417905	6.040369	16182.899521	6.040240	0.9938
	Roberts	16183.729781	6.040017	16186.627580	6.039239	0.9938
	LoG	16164.195769	6.045262	16165.060265	6.045030	0.9940
	Prewitt	16182.503474	6.040346	16182.868109	6.040248	0.9938
	Canny	16167.240296	6.044444	16163.095219	6.045558	0.9941
	Gabor	16361.034099	5.992696	16697.685714	5.904240	0.9933
	Fuzzy	16331.698492	6.000490	16668.438773	5.911854	0.9846
	Fractional conformable (1.16) $\alpha = 0.94$, $\beta = 0.91$ $\sigma = 1.45$	**11677.751458**	**7.457211**	**11533.201514**	**7.511304**	0.9942

Image Edge Detection Using Fractional Conformable Derivatives

FIGURE 1.4
Frequency analysis: (a) image with salt and pepper noise, (b) Fourier spectrum by a log transformation, (c) Fourier spectrum by a log transformation for $\alpha = 0.9$, $\beta = 0.9$, $\sigma = 1.45$, (d) Fourier spectrum by a log transformation for $\alpha = 0.5$, $\beta = 0.9$, $\sigma = 1.45$, (e) Fourier spectrum by a log transformation for $\alpha = 0.9$, $\beta = 0.5$, $\sigma = 1.45$, and (f) Fourier spectrum by a log transformation for $\alpha = 0.5$, $\beta = 0.5$, $\sigma = 1.45$.

and better than the Gabor method [46] and a fuzzy operator [47] because it has higher edge detection capability for both cases, strong and weak edges.

In order to analyse the effect of the fractional conformable operator on images with different kinds of noise, the Fourier spectrum is displayed in Figures 1.4 and 1.5. In these results, one can see that the fractional conformable operator is able to suppress high-frequency components (see Figures 1.4f and 1.5d), which it traduces in image smoothing. Also, depending of the value of the orders in this operator, the dominating directions could be attenuated. It can be seen easily in Figures 1.4d, f and 1.5d, f.

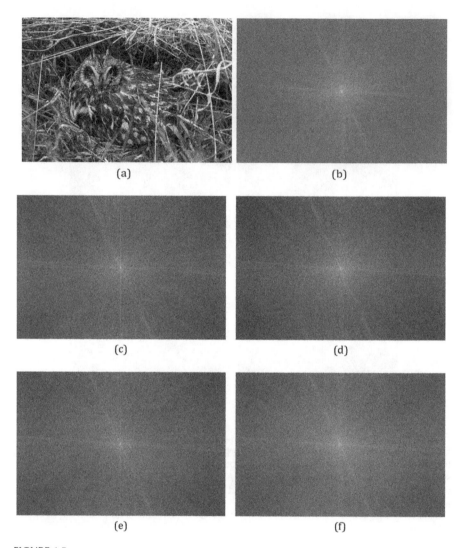

FIGURE 1.5
Frequency analysis: (a) image with Gaussian noise, (b) Fourier spectrum by a log transformation, (c) Fourier spectrum by a log transformation for $\alpha = 0.9$, $\beta = 0.9$, $\sigma = 1.45$, (d) Fourier spectrum by a log transformation for $\alpha = 0.5$, $\beta = 0.9$, $\sigma = 1.45$, (e) Fourier spectrum by a log transformation for $\alpha = 0.9$, $\beta = 0.5$, $\sigma = 1.45$, and (f) Fourier spectrum by a log transformation for $\alpha = 0.5$, $\beta = 0.5$, $\sigma = 1.45$.

Another interesting fact is that the high-frequency components in the image with salt and pepper noise are suppressed by lower α and β values, whereas in the image with Gaussian noise the same components are eliminated by a lower α value. On the other hand, the two orders involved in the fractional conformable operator provide us with an adaptive response because this operator is able to discriminate or not high- and low-frequency components. In image processing, it means that the proposed operator gets smooth parts, rough parts or both.

1.5 Test Images

The analysis presented in the previous section showed that the fractional conformable Gaussian gradient in Equation (1.16) has an important effect for detecting texture, which does not occur with traditional techniques. In this way, it is interesting to apply the proposed operator on medical images for extracting more structure feature details from the images that enable the identification of certain diseases or pathologies.

1.5.1 Cerebral Arteriovenous Malformation

Arteriovenous malformations (AVMs) are defects in the vascular system, consisting of tangles of abnormal blood vessels (nidus) in which the feeding arteries are directly connected to a venous drainage network without the interposition of a capillary bed [48]. An AVM can happen anywhere in the body, but brain AVMs present substantial risks when they bleed because they can rupture and bleed into the brain. Most AVMs can be detected with either a computed tomography (CT) brain scan or a magnetic resonance imaging (MRI) brain scan [49]. In this sense, it is interesting to implement the operator proposed in these types of medical images with the objective to identify potential AVMs. Figures 1.6 and 1.8 show MRIs of brains taken from [50] and [51], respectively; whereas those in Figures 1.7 and 1.9 show CTs taken from [52] and [53], respectively. All the preceding images are depicted without contrast enhancement. The experimental results show that the proposed operator improves the texture and contrast, thus achieving a more precise detection of AVMs.

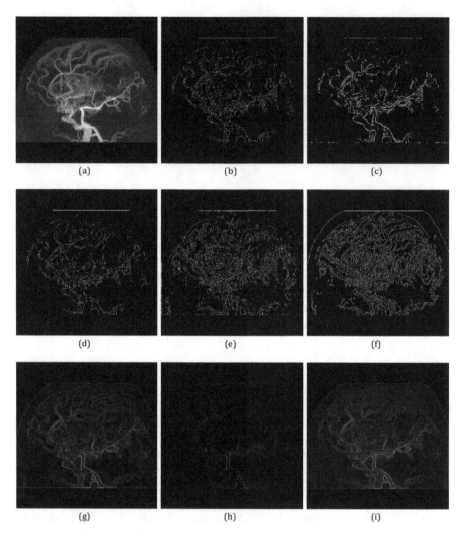

FIGURE 1.6
Cerebral arteriovenous malformation: (a) original image, (b) Sobel, (c) Roberts, (d) Prewitt, (e) LoG, (f) Canny, (g) Gabor, (h) Fuzzy, and (i) conformable fractional-order edge detector with $\alpha = 0.94$, $\beta = 0.91$, and $\sigma = 1.45$.

Image Edge Detection Using Fractional Conformable Derivatives 23

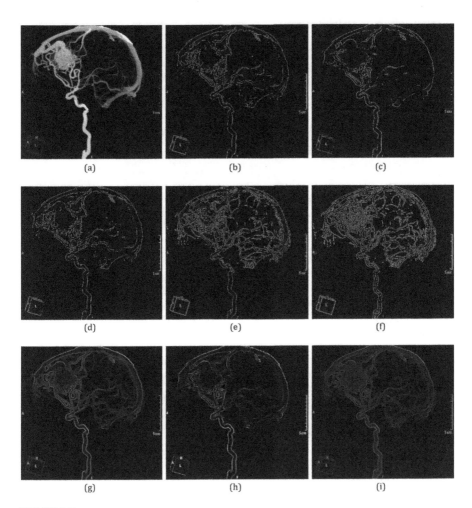

FIGURE 1.7
Cerebral arteriovenous malformation: (a) original image, (b) Sobel, (c) Roberts, (d) Prewitt, (e) LoG, (f) Canny, (g) Gabor, (h) Fuzzy, and (i) conformable fractional-order edge detector with $\alpha = 0.94$, $\beta = 0.91$, and $\sigma = 1.45$.

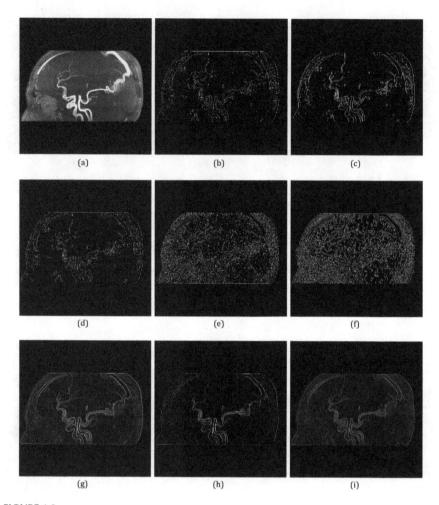

FIGURE 1.8
Cerebral arteriovenous malformation: (a) original image, (b) Sobel, (c) Roberts, (d) Prewitt, (e) LoG, (f) Canny, (g) Gabor, (h) Fuzzy, and (i) conformable fractional-order edge detector with $\alpha = 0.94$, $\beta = 0.91$, and $\sigma = 1.45$.

Image Edge Detection Using Fractional Conformable Derivatives   25

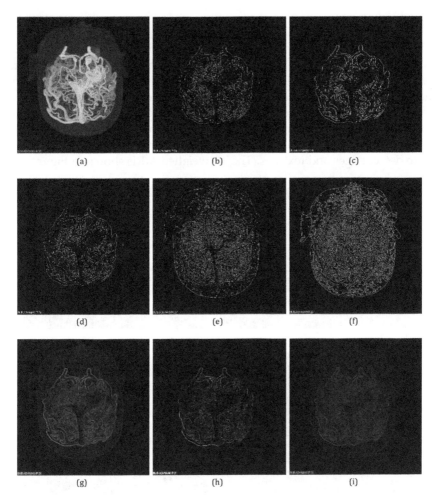

FIGURE 1.9
Cerebral arteriovenous malformation: (a) original image, (b) Sobel, (c) Roberts, (d) Prewitt, (e) LoG, (f) Canny, (g) Gabor, (h) Fuzzy, and (i) conformable fractional-order edge detector with $\alpha = 0.94$, $\beta = 0.91$, and $\sigma = 1.45$.

1.5.2 Meningioma

A meningioma is a tumour that arises from a layer of tissue (the meninges) that covers the brain and spine [54]. Although most meningiomas are encapsulated benign tumours with limited numbers of genetic aberrations, their intracranial location often leads to serious and potentially lethal consequences [55]. By using an MRI image, it is possible to determine the location of meningiomas; however, the detection and accurate diagnosis of meningiomas can be drastically improved with the use of the proposed operator due to its high capacity to detect edges and textures. The T1-weighed MRIs shown in Figures 1.10

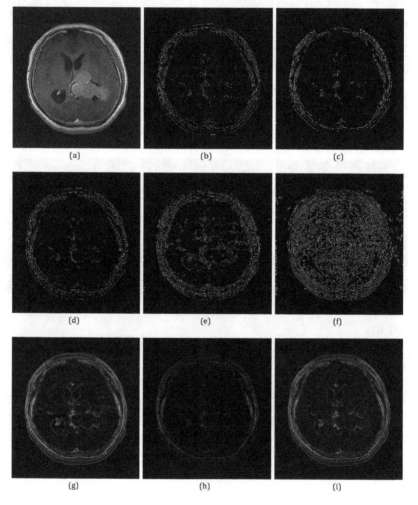

FIGURE 1.10
Intraventricular meningioma: (a) original image, (b) Sobel, (c) Roberts, (d) Prewitt, (e) LoG, (f) Canny, (g) Gabor, (h) Fuzzy, and (i) conformable fractional-order edge detector with $\alpha = 0.94$, $\beta = 0.91$, and $\sigma = 1.45$.

Image Edge Detection Using Fractional Conformable Derivatives

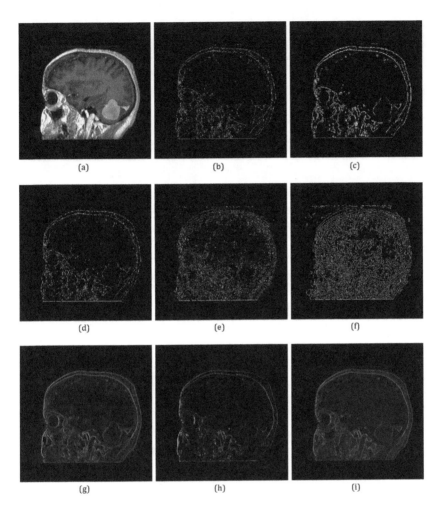

FIGURE 1.11
Tentorial meningioma: (a) Original image, (b) Sobel, (c) Roberts, (d) Prewitt, (e) LoG, (f) Canny, (g) Gabor, (h) Fuzzy, and (i) conformable fractional-order edge detector with $\alpha = 0.94$, $\beta = 0.91$, and $\sigma = 1.45$.

through 1.13 were taken from [56-59], respectively. In these figures, the boundaries of the meningiomas are correctly defined even in noisy images. By using the fractional conformable Gaussian edge, it is possible to make a better identification of the meningiomas achieving a more accurate diagnosis. On the other hand, with the boundaries of the meningiomas being well-defined, it is possible to obtain a better estimate of their size. In conclusion, the fractional conformable operator is able to identify much better the edges in MRI images for meningioma identification than classic operators such as Sobel, Prewitt, Roberts, LoG, and Canny.

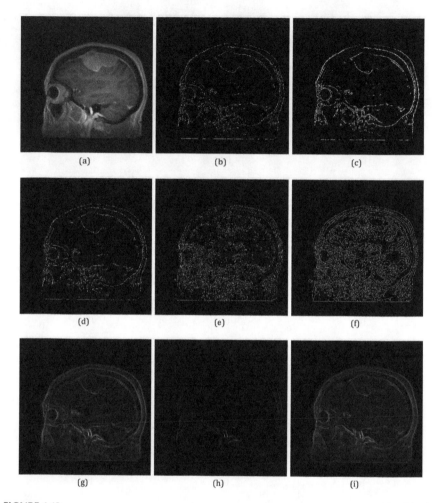

FIGURE 1.12
Rhabdoid meningioma: (a) original image, (b) Sobel, (c) Roberts, (d) Prewitt, (e) LoG, (f) Canny, (g) Gabor, (h) Fuzzy, and (i) conformable fractional-order edge detector with $\alpha = 0.94$, $\beta = 0.91$, and $\sigma = 1.45$.

Image Edge Detection Using Fractional Conformable Derivatives 29

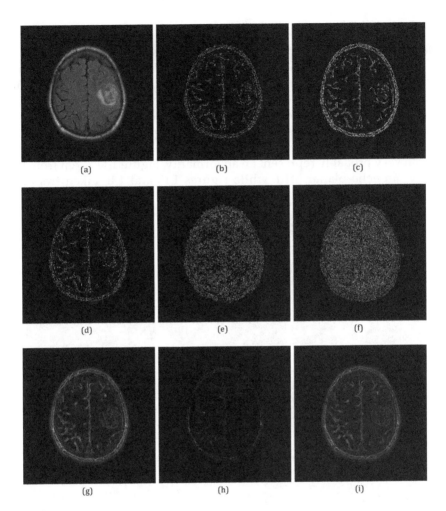

FIGURE 1.13
Microcystic meningioma: (a) original image, (b) Sobel, (c) Roberts, (d) Prewitt, (e) LoG, (f) Canny, (g) Gabor, (h) Fuzzy, and (i) conformable fractional-order edge detector with $\alpha = 0.94$, $\beta = 0.91$, and $\sigma = 1.45$.

1.5.3 Medulloblastoma

Medulloblastoma is a primary central nervous system tumour, that is, it begins in the brain or spinal cord [60]. It happens not only in children but also in adults; however, it is more common in children [61]. Medulloblastomas are very fast-growing; for this reason, early detection represents one of the most promising approaches to preserve life in patients. Edge detection has the potential to help in the early detection of medulloblastomas, by detecting changes in an MRI brain scan patterns, which very probably represent an abnormality. Figure 1.14 taken from [62] shows an echo-planar MRI, while Figures 1.15 and 1.16 taken from [63]

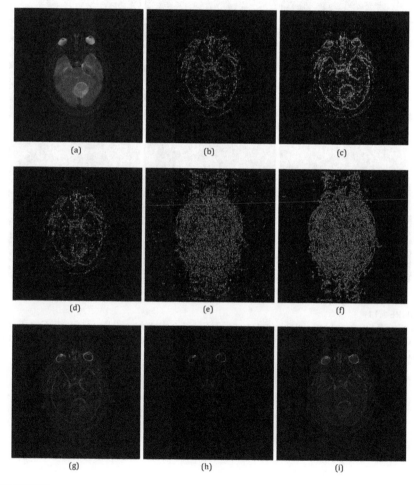

FIGURE 1.14

Medulloblastoma: (a) original image, (b) Sobel, (c) Roberts, (d) Prewitt, (e) LoG, (f) Canny, (g) Gabor, (h) Fuzzy, and (i) conformable fractional-order edge detector with $\alpha = 0.94$, $\beta = 0.91$, and $\sigma = 1.45$.

Image Edge Detection Using Fractional Conformable Derivatives　　　　31

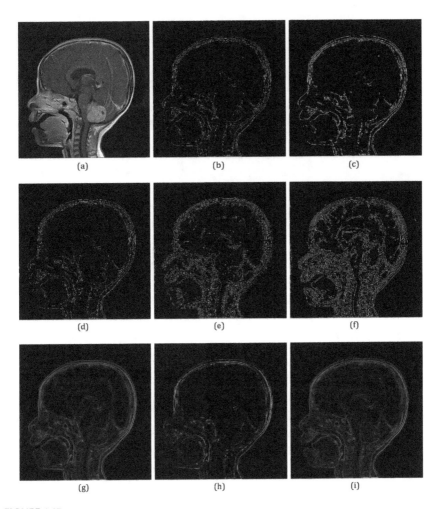

FIGURE 1.15
Medulloblastoma: (a) original image, (b) Sobel, (c) Roberts, (d) Prewitt, (e) LoG, (f) Canny, (g) Gabor, (h) Fuzzy, and (i) conformable fractional-order edge detector with $\alpha = 0.94$, $\beta = 0.91$, and $\sigma = 1.45$.

and [64], respectively, show T1-weighed MRIs. They show the performance comparisons between the proposed approach and the classical methods to find the abnormalities in the MRI brain scanners. According to these results, the proposed operator is more accurate for detecting the edges generated by the boundaries of the medulloblastomas; thus, their areas are well-defined.

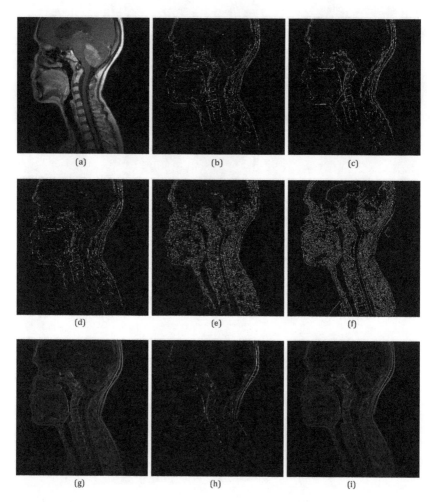

FIGURE 1.16
Medulloblastoma: (a) original image, (b) Sobel, (c) Roberts, (d) Prewitt, (e) LoG, (f) Canny, (g) Gabor, (h) Fuzzy, and (i) conformable fractional-order edge detector with $\alpha = 0.94$, $\beta = 0.91$, and $\sigma = 1.45$.

1.5.4 Abdominal Aortic Aneurysm

The aorta is the body's major blood vessel [65]. An abdominal aortic aneurysm (AAA) is a blood-filled bulge or ballooning in a part of your aorta that runs through the abdomen [66]. It usually causes no symptoms, except during rupture, which may result in pain in the abdomen or back, low blood

Image Edge Detection Using Fractional Conformable Derivatives 33

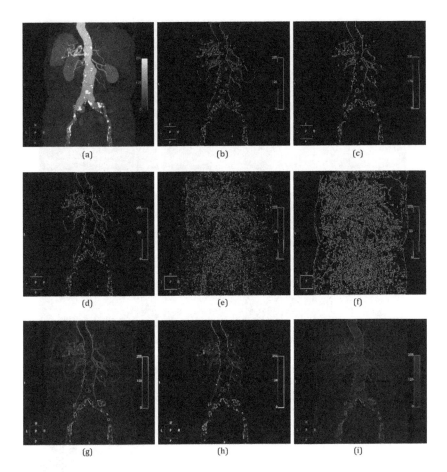

FIGURE 1.17
Abdominal aortic aneurysm: (a) original image, (b) Sobel, (c) Roberts, (d) Prewitt, (e) LoG, (f) Canny, (g) Gabor, (h) Fuzzy, and (i) conformable fractional-order edge detector with $\alpha = 0.94$, $\beta = 0.91$, and $\sigma = 1.45$.

pressure, or loss of consciousness, and often results in death [67]. The presence of an AAA can be confirmed with an abdominal ultrasound, abdominal and pelvic CT, or angiography. [68]. Figures 1.17 through 1.19 taken from [69–71], respectively show 3D maximum intensity projection (MIP) MRI images. These figures show that the proposed approach improves the detection and the segmentation of the aortic wall. This improvement represents a diagnostic system that allows analyses the abdominal MRI and warns if an abnormality is detected.

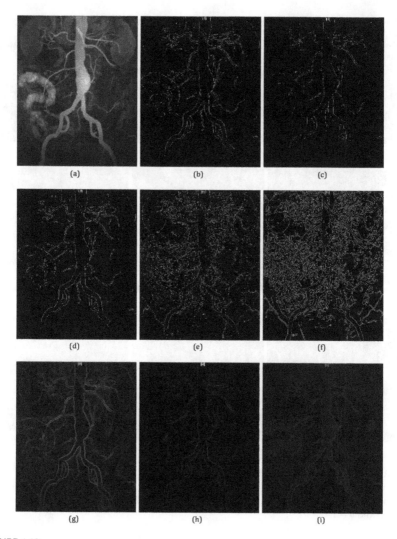

FIGURE 1.18
Abdominal aortic aneurysm: (a) original image, (b) Sobel, (c) Roberts, (d) Prewitt, (e) LoG, (f) Canny, (g) Gabor, (h) Fuzzy, and (i) conformable fractional-order edge detector with $\alpha = 0.94$, $\beta = 0.91$, and $\sigma = 1.45$.

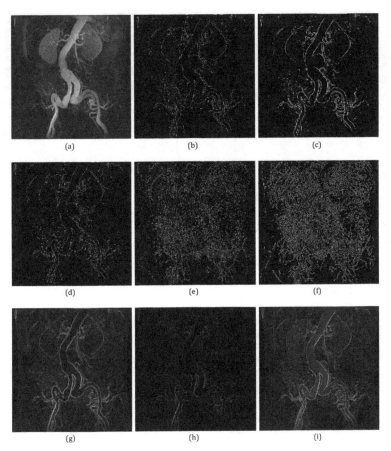

FIGURE 1.19
Abdominal aortic aneurysm: (a) original image, (b) Sobel, (c) Roberts, (d) Prewitt, (e) LoG, (f) Canny, (g) Gabor, (h) Fuzzy, and (i) conformable fractional-order edge detector with $\alpha = 0.94$, $\beta = 0.91$, and $\sigma = 1.45$.

1.5.5 Cerebral Venous Infarction

Cerebral venous infarctions are rare but serious conditions with devastating consequences without prompt diagnosis and treatment [72]. Symptoms of a cerebral infarction are determined by the parts of the brain affected [73]. For example, CT and MRI scanning will show damaged area in the brain [74]. Figures 1.20 and 1.22 taken from [75] and [76], respectively show T2-weighted MRI images of the brain, while Figure 1.21 shows a non-enhanced CT of the brain [77]. They also show that the fractional conformable derivative provides a rapid and a reliable method for cerebral venous infarction detection, since it identifies the parts of the brain affected.

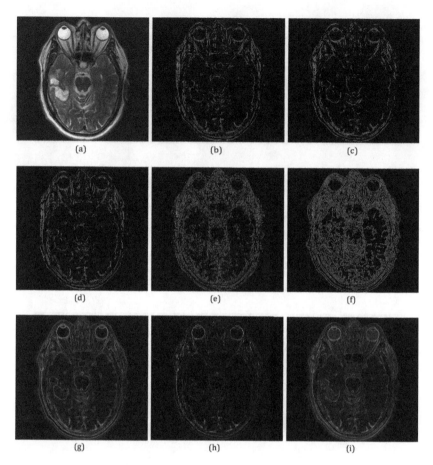

FIGURE 1.20
Cerebral venous infarction: (a) original image, (b) Sobel, (c) Roberts, (d) Prewitt, (e) LoG, (f) Canny, (g) Gabor, (h) Fuzzy, and (i) conformable fractional-order edge detector with $\alpha = 0.94$, $\beta = 0.91$, and $\sigma = 1.45$.

Image Edge Detection Using Fractional Conformable Derivatives

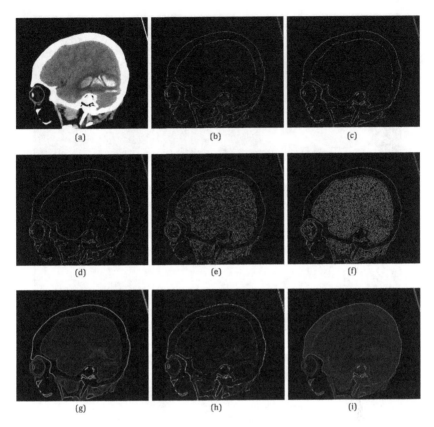

FIGURE 1.21
Cerebral venous hemorrhagic infarction: (a) original image, (b) Sobel, (c) Roberts, (d) Prewitt, (e) LoG, (f) Canny, (g) Gabor, (h) Fuzzy, and (i) conformable fractional-order edge detector with $\alpha = 0.94$, $\beta = 0.91$ and $\sigma = 1.45$.

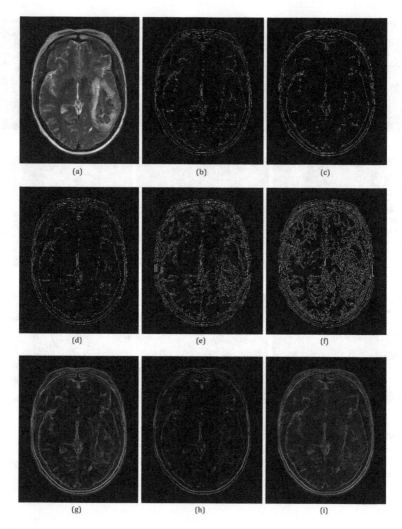

FIGURE 1.22
Cerebral venous hemorrhagic infarction: (a) original image, (b) Sobel, (c) Roberts, (d) Prewitt, (e) LoG, (f) Canny, (g) Gabor, (h) Fuzzy, and (i) conformable fractional-order edge detector with $\alpha = 0.94$, $\beta = 0.91$ and $\sigma = 1.45$.

1.5.6 Breast Calcifications

Breast calcifications are small calcium deposits that develop in a woman's breast tissue [78]. They are very common and usually benign (non-cancerous); however, certain kinds of them may suggest early symptoms of breast cancer if it appears as a small-size bright spot in a mammogram image [79]. In order to prevent breast cancer, a timely diagnosis is necessary. In this sense, the proposed operator represents a useful tool for the diagnosis and treatment of breast cancer. The mammogram images (Figures 1.23 through 1.25) were taken from [80–82], respectively. They show that the proposed operator improves image contrast and texture of the mammogram images, making it easier to detect calcifications.

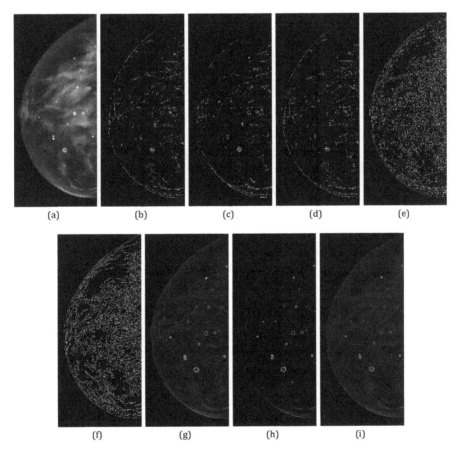

FIGURE 1.23
Breast calcifications: (a) original image, (b) Sobel, (c) Roberts, (d) Prewitt, (e) LoG, (f) Canny, (g) Gabor, (h) Fuzzy, and (i) conformable fractional-order edge detector with $\alpha = 0.94$, $\beta = 0.91$ and $\sigma = 1.45$.

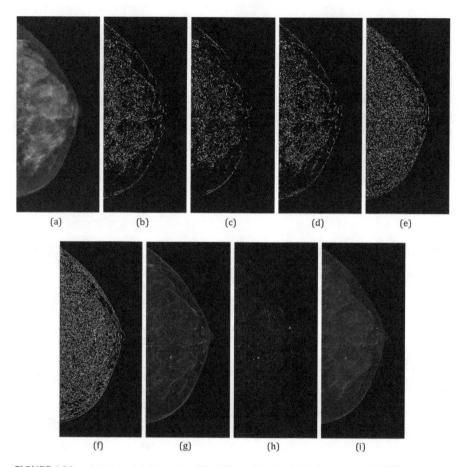

FIGURE 1.24
Breast calcifications: (a) original image, (b) Sobel, (c) Roberts, (d) Prewitt, (e) LoG, (f) Canny, (g) Gabor, (h) Fuzzy, and (i) conformable fractional-order edge detector with $\alpha = 0.94$, $\beta = 0.91$, and $\sigma = 1.45$.

Experimental results show that the proposed fractional conformable operator yields better visual effects than other traditional techniques. In addition, the calculated values of MSE and PSNR, using Equations (1.17) and (1.18) for the medical images (see Tables 1.8 through 1.13) show that the proposed approach has higher PSNR, which means that it is remarkably robust to the noise produced in the image formation process and the noise due to

Image Edge Detection Using Fractional Conformable Derivatives

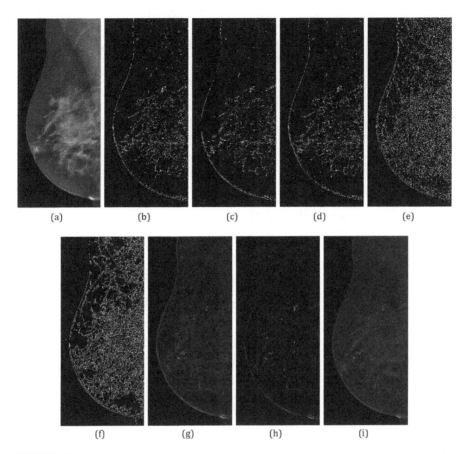

FIGURE 1.25
Breast calcifications: (a) original image, (b) Sobel, (c) Roberts, (d) Prewitt, (e) LoG, (f) Canny, (g) Gabor, (h) Fuzzy, and (i) conformable fractional-order edge detector with $\alpha = 0.94$, $\beta = 0.91$, and $\sigma = 1.45$.

illumination changes of the scene. To sum up, it can be said that the proposed fractional conformable edge detector has a better edge detection capability even in the presence of noise on different kinds of medical images like a CT brain scan or a MRI brain scan. Thus, the proposed approach represents an important tool in medical image visualisation, localisation of pathology, and computer-aided diagnosis.

TABLE 1.8

Performance Parameters of Cerebral Arteriovenous Malformation Images

	Operator	MSE	PSNR (dB)	Image
Cerebral arteriovenous malformation	Sobel	3783.126682	12.352294	
	Roberts	3782.778001	12.352695	
	LoG	3780.507876	12.355302	
	Prewitt	3783.157542	12.352259	
	Canny	3778.545011	12.357557	
	Gabor	3785.894319	12.349118	
	Fuzzy	3786.043471	12.348947	
	Fractional conformable (1.16) $\alpha = 0.94$, $\beta = 0.91$, $\sigma = 145$	**3300.764715**	**12.944657**	
	Sobel	3416.662774	12.794782	
	Roberts	3416.733039	12.794693	
	LoG	3414.556140	12.797461	
	Prewitt	3416.670268	12.794772	
	Canny	3412.912478	12.799552	
	Gabor	3418.255584	12.792758	
	Fuzzy	3417.107271	12.794217	
	Fractional conformable (1.16) $\alpha = 0.94$, $\beta = 0.91$ $\sigma = 1.45$	**2895.540405**	**13.513507**	
	Sobel	5020.748638	11.123118	
	Roberts	5020.274991	11.123528	
	LoG	5015.270126	11.127860	
	Prewitt	5020.765995	11.123103	
	Canny	5012.355637	11.130384	
	Gabor	5021.778099	11.122228	
	Fuzzy	5021.912650	11.122112	
	Fractional conformable (1.16) $\alpha = 0.94$, $\beta = 0.91$ $\sigma = 1.45$	**4312.946957**	**11.783062**	
	Sobel	5755.337341	10.530095	
	Roberts	5755.842567	10.529714	
	LoG	5750.278827	10.533914	
	Prewitt	5755.318443	10.530110	
	Canny	5747.264440	10.536191	
	Gabor	5758.851799	10.527444	
	Fuzzy	5758.214968	10.527924	
	Fractional conformable (1.16) $\alpha = 0.94$, $\beta = 0.91$, $\sigma = 1.45$	**4943.290527**	**11.190642**	

Image Edge Detection Using Fractional Conformable Derivatives 43

TABLE 1.9

Performance Parameters of Meningioma Images

	Operator	MSE	PSNR (dB)	Image
Meningioma	Sobel	5143.767475	11.017990	
	Roberts	5143.185473	11.018481	
	LoG	5141.253708	11.020113	
	Prewitt	5143.782814	11.017977	
	Canny	5135.084165	11.025327	
	Gabor	5142.667260	11.018919	
	Fuzzy	5146.417567	11.015753	
	Fractional conformable (1.16) $\alpha = 0.94, \beta = 0.91, \sigma = 1.45$	**4455.874713**	**11.641473**	
	Sobel	7258.737987	9.522192	
	Roberts	7258.253780	9.522482	
	LoG	7255.606486	9.524066	
	Prewitt	7258.762060	9.522178	
	Canny	7250.722598	9.526990	
	Gabor	7261.963603	9.520262	
	Fuzzy	7261.603144	9.520478	
	Fractional conformable (1.16) $\alpha = 0.94, \beta = 0.91, \sigma = 1.45$	**6326.101909**	**10.119441**	
	Sobel	3725.129704	12.419389	
	Roberts	3724.747217	12.419835	
	LoG	3722.623518	12.422312	
	Prewitt	3725.153825	12.419361	
	Canny	3719.860110	12.425537	
	Gabor	3727.303354	12.416856	
	Fuzzy	3727.492989	12.416635	
	Fractional conformable (1.16) $\alpha = 0.94, \beta = 0.91, \sigma = 1.45$	**3313.022053**	**12.928560**	
	Sobel	2111.980658	14.883904	
	Roberts	2111.494174	14.884904	
	LoG	2108.556831	14.890950	
	Prewitt	2111.994381	14.883876	
	Canny	2106.775678	14.894620	
	Gabor	2113.449987	14.880883	
	Fuzzy	2114.191716	14.879359	
	Fractional conformable (1.16) $\alpha = 0.94, \beta = 0.91, \sigma = 1.45$	**1834.168109**	**15.496412**	

TABLE 1.10

Performance Parameters of Medulloblastoma Images

	Operator	MSE	PSNR (dB)	Image
Medulloblastoma	Sobel	835.214440	18.912823	
	Roberts	834.9312	18.914296	
	LoG	833.275136	18.922919	
	Prewitt	835.213390	18.912829	
	Canny	832.748676	18.925664	
	Gabor	836.478411	18.906256	
	Fuzzy	836.524607	18.906016	
	Fractional conformable (1.16) $\alpha = 0.94$, $\beta = 0.91$, $\sigma = 1.45$	**723.746542**	**19.534938**	
	Sobel	5263.997623	10.917646	
	Roberts	5263.498474	10.918058	
	LoG	5260.480026	10.920549	
	Prewitt	5264.013243	10.917633	
	Canny	5257.117500	10.923326	
	Gabor	5266.662369	10.915448	
	Fuzzy	5264.969212	10.916845	
	Fractional conformable (1.16) $\alpha = 0.94$, $\beta = 0.91$, $\sigma = 1.45$	**4500.063957**	**11.598616**	
	Sobel	3581.192315	12.590527	
	Roberts	3580.960999	12.590807	
	LoG	3577.159834	12.595420	
	Prewitt	3581.202743	12.590514	
	Canny	3575.598143	12.597316	
	Gabor	3583.004418	12.588330	
	Fuzzy	3582.552275	12.588878	
	Fractional conformable (1.16) $\alpha = 0.94$, $\beta = 0.91$, $\sigma = 1.45$	**3073.448542**	**13.254544**	

Image Edge Detection Using Fractional Conformable Derivatives 45

TABLE 1.11

Performance Parameters of Abdominal Aortic Aneurysm Images

	Operator	MSE	PSNR (dB)	Image
Abdominal aortic aneurysm	Sobel	4085.195184	12.018675	
	Roberts	4085.034403	12.018846	
	LoG	4081.068547	12.023064	
	Prewitt	4085.191186	12.018679	
	Canny	4078.028095	12.026301	
	Gabor	4086.572231	12.017211	
	Fuzzy	4085.159588	12.018713	
	Fractional conformable (1.16) $\alpha = 0.94$, $\beta = 0.91$, $\sigma = 1.45$	**3476.175309**	**12.719786**	
	Sobel	5789.123507	10.504675	
	Roberts	5789.147733	10.504657	
	LoG	5785.698733	10.507245	
	Prewitt	5789.134011	10.504667	
	Canny	5779.605415	10.511821	
	Gabor	5792.054816	10.502476	
	Fuzzy	5791.955795	10.502551	
	Fractional conformable (1.16) $\alpha = 0.94$, $\beta = 0.91$, $\sigma = 1.45$	**5080.259647**	**11.071944**	
	Sobel	4521.974876	11.577522	
	Roberts	4521.522835	11.577956	
	LoG	4517.260085	11.582052	
	Prewitt	4521.993122	11.577504	
	Canny	4512.407279	11.586720	
	Gabor	4522.515444	11.577003	
	Fuzzy	4524.523092	11.575075	
	Fractional conformable (1.16) $\alpha = 0.94$, $\beta = 0.91$, $\sigma = 1.45$	**3938.006367**	**12.178039**	

TABLE 1.12
Performance Parameters of Cerebral Venous Infarction Images

	Operator	MSE	PSNR (dB)	Image
Cerebral venous infarction	Sobel	6674.994753	9.886294	
	Roberts	6675.211790	9.886153	
	LoG	6670.894863	9.888962	
	Prewitt	6674.988833	9.886298	
	Canny	6667.026803	9.891481	
	Gabor	6678.344247	9.884115	
	Fuzzy	6678.335675	9.884121	
	Fractional conformable (1.16) $\alpha = 0.94, \beta = 0.91$ $\sigma = 1.45$	**5684.050580**	**10.584224**	
	Sobel	14368.117728	6.556804	
	Roberts	14367.894657	6.556872	
	LoG	14361.467700	6.558815	
	Prewitt	14368.116982	6.556805	
	Canny	14356.650301	6.560272	
	Gabor	14367.419615	6.557015	
	Fuzzy	14366.940537	6.557160	
	Fractional conformable (1.16) $\alpha = 0.94, \beta = 0.91,$ $\sigma = 1.45$	**12794.205730**	**7.060670**	
	Sobel	5473.124770	10.748450	
	Roberts	5472.658648	10.748820	
	LoG	5469.447497	10.751369	
	Prewitt	5473.136758	10.748440	
	Canny	5465.814620	10.754254	
	Gabor	5473.416432	10.748218	
	Fuzzy	5475.269908	10.746748	
	Fractional conformable (1.16) $\alpha = 0.94, \beta = 0.91,$ $\sigma = 1.45$	**4676.396780**	**11.431690**	

Image Edge Detection Using Fractional Conformable Derivatives 47

TABLE 1.13

Performance Parameters of Breast Calcifications Images

	Operator	MSE	PSNR (dB)	Image
Breast calcifications	Sobel	3664.116021	12.491111	
	Roberts	3664.625012	12.490508	
	LoG	3656.511288	12.500134	
	Prewitt	3664.129940	12.491094	
	Canny	3654.725622	12.502255	
	Gabor	3666.355867	12.488457	
	Fuzzy	3665.904068	12.488992	
	Fractional conformable (1.16) $\alpha = 0.94$, $\beta = 0.91$, $\sigma = 1.45$	**3163.445956**	**13.129199**	
	Sobel	3496.811684	12.694081	
	Roberts	3498.448598	12.692048	
	LoG	3492.086278	12.699953	
	Prewitt	3497.008547	12.693836	
	Canny	3489.425411	12.703264	
	Gabor	3504.440257	12.684617	
	Fuzzy	3504.351619	12.684726	
	Fractional conformable (1.16) $\alpha = 0.94$, $\beta = 0.91$, $\sigma = 1.45$	**3014.228161**	**13.339042**	
	Sobel	2928.975706	13.463645	
	Roberts	2929.577087	13.462754	
	LoG	2922.894874	13.472671	
	Prewitt	2929.016962	13.463584	
	Canny	2919.890927	13.477137	
	Gabor	2932.662068	13.458183	
	Fuzzy	2932.394604	13.458579	
	Fractional conformable (1.16) $\alpha = 0.94$, $\beta = 0.91$, $\sigma = 1.45$	**2522.718155**	**14.112116**	

TABLE 1.14

Execution Time Results by Using All Medical Images Shown in This Proposal

Execution Time	MATLAB	GNU Octave
Min	0.015262 [s]	0.41500 [s]
Max	0.039825 [s]	0.52244 [s]
Mean	0.025038 [s]	0.45644 [s]

TABLE 1.15

Execution Time of the Proposed Method by Using a High-Resolution Image in Different Programming Languages

	MATLAB	GNU Octave
Width	3504	3504
Height	2336	2336
Execution time	0.546291 [s]	1.90993 [s]

To show the computational effort of the proposed method, it was programmed in different languages. Also, this approach was tested by using high-resolution images. These results are shown in Tables 1.14 and 1.15, and all of them were carried out with computational characteristics of: AMD A12-9720P @2.7GHz, 8GB RAM, GNU Archlinux with KDE Plasma 5.16.3.

1.6 Discussion and Conclusion

A novel edge detector based on a fractional conformable derivative was developed in this work for processing and analysing medical images, such as CT brain scans and MRI brain scans. The analysis showed better effectiveness of the proposed operator to detect both strong edges (boundaries of objects) and weak edges (texture) over some classical existing methods, such as Sobel, Roberts, LoG, Prewitt, and Canny. According to this comparative analysis, it was also demonstrated that the proposed fractional conformable operator is robust to noise. This operator has been implemented for edge detection on CT brain scans and MRI brain scans to identify the complicated structure of cerebral AVMs, meningiomas, and medulloblastomas. The results led to the conclusion that the proposed approach not only maintains the low-frequency contour features in the smooth areas of the image, but they also enhance edges and textures corresponding to high-frequency image components. A quantitative analysis was conducted to verify that the use of the proposed edge detector generates a higher PSNR than the existing methods, which implies a superior texture enhancement in medical images. The optimal choice of the conformable order and the fractional order depend on the type of images considered. In conclusion, the proposed edge detector depicts potential as an early predictor of disease onset and prognosis of cerebral AVMs, meningiomas, and medulloblastomas. However, our proposed technique is not only limited to medical images but can also be applied to analyse different kind of images, such as satellite images and images taken by drones, among others.

Acknowledgements

Jorge Enrique Lavín Delgado and Jesús Emmanuel Solís Pérez acknowledges the support provided by CONACyT through the assignment post-doctoral and doctoral fellowship, respectively. José Francisco Gómez Aguilar acknowledges the support provided by CONACyT: Cátedras CONACyT para jóvenes investigadores 2014. José Francisco Gómez Aguilar and Ricardo Fabricio Escobar Jiménez acknowledge the support provided by SNI-CONACyT.

Competing Interests

The authors declare that there is no conflict of interest regarding the publication of this paper.

Authors' Contributions

All authors contributed equally and significantly in writing this article. All authors read and approved the final manuscript.

References

1. R. Anas, H. A. Elhadi, E. S. Ali, Impact of edge detection algorithms in medical image processing, *World Scientific News* 118 (2019) 129–143.
2. B. Dhruv, N. Mittal, M. Modi, Comparative analysis of edge detection techniques for medical images of different body parts, in: *International Conference on Recent Developments in Science, Engineering and Technology*, Springer, 2017, pp. 164–176.
3. B. van Ginneken, S. Kerkstra, J. Meakin, Grand challenges in biomedical image analysis (2018). https://grand-challenge.org/challenges/.
4. D. Ziou, S. Tabbone, Edge detection techniques-an overview, *Pattern Recognition and Image Analysis C/C of Raspoznavaniye Obrazov I Analiz Izobrazhenii* 8 (1998) 537–559.
5. L. G. Roberts, Machine perception of three-dimensional solids, PhD thesis, Massachusetts Institute of Technology (1963).
6. J. M. Prewitt, Object enhancement and extraction, *Picture Processing and Psychopictorics* 10 (1) (1970) 15–19.
7. I. Sobel, "History and definition of the sobel operator," Retrieved from the World Wide Web, 2014.
8. J. Canny, A computational approach to edge detection, in: *Readings in Computer Vision*, Elsevier, 1987, pp. 184–203.
9. C.-C. Chen, Fast boundary detection: A generalization and a new algorithm, *IEEE Transactions on Computers* 100 (10) (1977) 988–998.
10. D. Marr, E. Hildreth, Theory of edge detection, *Proceedings of the Royal Society of London. Series B. Biological Sciences* 207 (1167) (1980) 187–217.
11. R. M. Haralick, Digital step edges from zero crossing of second directional derivatives, in: *Readings in Computer Vision*, Elsevier, 1987, pp. 216–226.
12. K. Oldham, J. Spanier, *The Fractional Calculus Theory and Applications of Differentiation and Integration to Arbitrary Order*, Vol. 111, Elsevier, 1974.
13. Q. Yang, D. Chen, T. Zhao, Y. Chen, Fractional calculus in image processing: A review, *Fractional Calculus and Applied Analysis* 19 (5) (2016) 1222–1249.
14. S.-C. Pei, M.-H. Yeh, Discrete fractional Hilbert transform, *IEEE Transactions on Circuits and Systems II: Analog and Digital Signal Processing* 47 (11) (2000) 1307–1311.
15. B. Mathieu, P. Melchior, A. Oustaloup, C. Ceyral, Fractional differentiation for edge detection, *Signal Processing* 83 (11) (2003) 2421–2432.
16. H. Yang, Y. Ye, D. Wang, B. Jiang, A novel fractional-order signal processing based edge detection method, in: *2010 11th International Conference on Control Automation Robotics & Vision*, IEEE, 2010, pp. 1122–1127.
17. Y. Zhang, Y. Pu, J. Zhou, Construction of fractional differential masks based on Riemann-Liouville definition, *Journal of Computational Information Systems* 6 (10) (2010) 3191–3199.
18. X. Pan, Y. Ye, J. Wang, X. Gao, Novel fractional-order calculus masks and compound derivatives with applications to edge detection, in: *2013 6th International Congress on Image and Signal Processing (CISP)*, Vol. 1, IEEE, 2013, pp. 309–314.
19. D. Tian, J. Wu, Y. Yang, A fractional-order edge detection operator for medical image structure feature extraction, in: *The 26th Chinese Control and Decision Conference (2014 CCDC)*, IEEE, 2014, pp. 5173–5176.

20. C. Telke, M. Beitelschmidt, Edge detection based on fractional order differentiation and its application to railway track images, *PAMM* 15 (1) (2015) 671–672.
21. M. Hacini, A. Hacini, H. Akdag, F. Hachouf, A 2D-fractional derivative mask for image feature edge detection, in: *2017 International Conference on Advanced Technologies for Signal and Image Processing (ATSIP)*, IEEE, 2017, pp. 1–6.
22. S. Hemalatha, S. M. Anouncia, A computational model for texture analysis in images with fractional differential filter for texture detection, in: *Biometrics: Concepts, Methodologies, Tools, and Applications*, IGI Global, 2017, pp. 309–332.
23. S. Kumar, R. Saxena, K. Singh, Fractional Fourier transform and fractional-order calculus-based image edge detection, *Circuits, Systems, and Signal Processing* 36 (4) (2017) 1493–1513.
24. A. Nandal, H. Gamboa-Rosales, A. Dhaka, J. M. Celaya-Padilla, J. I. Galvan-Tejada, C. E. Galvan-Tejada, F. J. Martinez-Ruiz, C. Guzman-Valdivia, Image edge detection using fractional calculus with feature and contrast enhancement, *Circuits, Systems, and Signal Processing* 1 (2018) 1–27.
25. W. S. ElAraby, A. H. Madian, M. A. Ashour, I. Farag, M. Nassef, Fractional edge detection based on genetic algorithm, in: *2017 29th International Conference on Microelectronics (ICM)*, IEEE, 2017, pp. 1–4.
26. S. Gupta, A. Bhardwaj, R. Sharma, P. Varshney, S. Srivastava, Image edge detection using fractional order differential calculus, in: *Proceedings of the 2018 the 2nd International Conference on Video and Image Processing*, ACM, 2018, pp. 44–48.
27. S. Godbole, G. Phadke, Satellite image edge detection using fractional order method, in: *2018 IEEE 13th International Conference on Industrial and Information Systems (ICIIS)*, IEEE, 2018, pp. 130–135.
28. J. Reju, N. Kunju, Detection of Alzheimer's disease using fractional edge detection, *Global Journal of Technology and Optimization* 9 (2018) 230.
29. H. Jalalinejad, A. Tavakoli, F. Zarmehi, A simple and flexible modification of Grünwald–Letnikov fractional derivative in image processing, *Mathematical Sciences* 12 (3) (2018) 205–210.
30. T. Bento, D. Valério, P. Teodoro, J. Martins, Fractional order image processing of medical images, *Journal of Applied Nonlinear Dynamics* 6 (2) (2017) 181–191.
31. I. Podlubny, *Fractional Differential Equations: An Introduction to Fractional Derivatives, Fractional Differential Equations, to Methods of Their Solution and Some of Their Applications*, Vol. 198, Elsevier, 1998.
32. R. Agarwal, S. D. Purohit, A mathematical fractional model with nonsingular kernel for thrombin receptor activation in calcium signalling, *Mathematical Methods in the Applied Sciences* 1 (1) (2019) 1–16.
33. D. Kumar, J. Singh, S. D. Purohit, R. Swroop, A hybrid analytical algorithm for nonlinear fractional wave-like equations, *Mathematical Modelling of Natural Phenomena* 14 (3) (2019) 1–11.
34. J. Sharmaa, K. Sharma, S. D. Purohit, A. Atangana, Hybrid watermarking algorithm using finite radon and fractional Fourier transform, *Fundamental Informaticae* 151 (1–4) (2017) 523–543.
35. R. Khalil, M. Al Horani, A. Yousef, M. Sababheh, A new definition of fractional derivative, *Journal of Computational and Applied Mathematics* 264 (2014) 65–70.
36. F. Jarad, E. Uğurlu, T. Abdeljawad, D. Baleanu, On a new class of fractional operators, *Advances in Difference Equations* 2017 (1) (2017) 247.

37. V. E. Tarasov, Differential equations with fractional derivative and universal map with memory, *Journal of Physics A: Mathematical and Theoretical* 42 (46) (2009) 465102.
38. J. Pérez, J. Gómez-Aguilar, D. Baleanu, F. Tchier, Chaotic attractors with fractional conformable derivatives in the Liouville–Caputo sense and its dynamical behaviors, *Entropy* 20 (5) (2018) 384.
39. D. Sadykova, A. P. James, Quality assessment metrics for edge detection and edge-aware filtering: A tutorial review, in: *2017 International Conference on Advances in Computing, Communications and Informatics (ICACCI)*, IEEE, 2017, pp. 2366–2369.
40. Q. Huynh-Thu, M. Ghanbari, Scope of validity of PSNR in image/video quality assessment, *Electronics Letters* 44 (13) (2008) 800–801.
41. J. López-Randulfe, C. Veiga, J. J. Rodríguez-Andina, J. Farina, A quantitative method for selecting denoising filters, based on a new edge-sensitive metric, in: *2017 IEEE International Conference on Industrial Technology (ICIT)*, IEEE, 2017, pp. 974–979.
42. Z. Wang, A. C. Bovik, H. R. Sheikh, E. P. Simoncelli, et al., Image quality assessment: From error visibility to structural similarity, *IEEE Transactions on Image Processing* 13 (4) (2004) 600–612.
43. G.-H. Chen, C.-L. Yang, L.-M. Po, S.-L. Xie, Edge-based structural similarity for image quality assessment, in: *2006 IEEE International Conference on Acoustics Speech and Signal Processing Proceedings*, Vol. 2, IEEE, 2006, pp. 933–936.
44. B. Jähne, H. Haussecker, P. Geissler, *Handbook of Computer Vision and Applications*, Vol. 2, Citeseer, 1999.
45. D. Martin, C. Fowlkes, D. Tal, J. Malik, et al., A database of human segmented natural images and its application to evaluating segmentation algorithms and measuring ecological statistics, ICCV, Vancouver, 2001.
46. R. Mehrotra, K. R. Namuduri, N. Ranganathan, Gabor filter-based edge detection, *Pattern Recognition* 25 (12) (1992) 1479–1494.
47. C. J. Miosso, A. Bauchspiess, Fuzzy inference system applied to edge detection in digital images, in: *Proceedings of the V Brazilian Conference on Neural Networks*, 2001, pp. 481–486.
48. J. van Beijnum, H. B. van der Worp, D. R. Buis, R. A.-S. Salman, L. J. Kappelle, G. J. Rinkel, J. W. B. van der Sprenkel, W. P. Vandertop, A. Algra, C. J. Klijn, Treatment of brain arteriovenous malformations: A systematic review and meta-analysis, *JAMA* 306 (18).
49. C. S. Ogilvy, P. E. Stieg, I. Awad, R. D. Brown Jr, D. Kondziolka, R. Rosenwasser, W. L. Young, G. Hademenos, Recommendations for the management of intracranial arteriovenous malformations: A statement for healthcare professionals from a special writing group of the stroke council, American Stroke Association, *Circulation* 103 (21) (2001) 2644–2657.
50. F. Gaillard, Radiopaedia, see https://radiopaedia.org/cases/cerebral-arteriovenous-malformation-huge?lang=us (2016).
51. K. Arunkumar Mistry, Radiopaedia, see https://radiopaedia.org/cases/cerebral-arteriovenous- malformation-12?lang=us (2015).
52. D. Ibrahim, Radiopaedia, see https://radiopaedia.org/cases/cerebral-arteriovenous-malformation-8?lang=us (2014).
53. A. Goel, Radiopaedia, see https://radiopaedia.org/cases/cerebral-arteriovenous-malformation-7?lang=us (2018).

54. C. Marosi, M. Hassler, K. Roessler, M. Reni, M. Sant, E. Mazza, C. Vecht, Meningioma, *Critical Reviews in Oncology/Hematology* 67 (2) (2008) 153–171.
55. J. Wiemels, M. Wrensch, E. B. Claus, Epidemiology and etiology of meningioma, *Journal of Neurooncology* 99 (3) (2010) 307–314.
56. I. Bickle, Radiopaedia, see https://radiopaedia.org/cases/intraventricular-meningioma-7?lang=us (2016).
57. H. Salam, Radiopaedia, see https://radiopaedia.org/cases/tentorial-meningioma?lang=us (2010).
58. B. Di Muzio, Radiopaedia, see https://radiopaedia.org/cases/rhabdoid-meningioma-2?lang=us (2016).
59. F. Gaillard, Radiopaedia, see https://radiopaedia.org/cases/microcystic-meningioma-1?lang=us (2017).
60. C. Prazaru, G. Diaconu, A. Ulinici, H. A. Ainain, A. S. Prazaru, I. Grigore, Medulloblastoma-clinical and therapeutical aspects, *Romanian Journal of Pediatrics* 66 (2).
61. A. A. Friedman, A. Letai, D. E. Fisher, K. T. Flaherty, Precision medicine for cancer with next-generation functional diagnostics, *Nature Reviews Cancer* 15 (12) (2015) 747.
62. F. Gaillard, Radiopaedia, see https://radiopaedia.org/cases/medulloblastoma-22?lang=us (2014).
63. S. Soltany Hosn, Radiopaedia, see https://radiopaedia.org/cases/medulloblastoma-18?lang=us (2013).
64. A. Al Khateeb, Radiopaedia, see https://radiopaedia.org/cases/medulloblastoma-38?lang=us (2016).
65. N. Sakalihasan, R. Limet, O. D. Defawe, Abdominal aortic aneurysm, *The Lancet* 365 (9470) (2005) 1577–1589.
66. M. P. Nevitt, D. J. Ballard, J. W. Hallett Jr, Prognosis of abdominal aortic aneurysms, *New England Journal of Medicine* 321 (15) (1989) 1009–1014.
67. C. B. Ernst, Abdominal aortic aneurysm, *New England Journal of Medicine* 328 (16) (1993) 1167–1172.
68. G. H. White, W. Yu, J. May, X. Chaufour, M. S. Stephen, Endoleak as a complication of endoluminal grafting of abdominal aortic aneurysms: Classification, incidence, diagnosis, and management, *Journal of Endovascular Therapy* 4 (2) (1997) 152–168.
69. D. Cuete, Radiopaedia, see https://radiopaedia.org/cases/abdominal-aortic-aneurysm-25?lang=us (2015).
70. R. Schubert, Radiopaedia, see https://radiopaedia.org/cases/abdominal-aortic-aneurysm-mra- mri?lang=us (2015).
71. R. Schubert, Radiopaedia, see https://radiopaedia.org/cases/dilative-arteriopathy-of-the-abdominal-aorta-and-iliac-arteries?lang=us (2015).
72. J. L. Leach, R. B. Fortuna, B. V. Jones, M. F. Gaskill-Shipley, Imaging of cerebral venous thrombosis: Current techniques, spectrum of findings, and diagnostic pitfalls, *Radiographics* 26 (suppl_1) (2006) S19–S41.
73. C. Bassetti, M. S. Aldrich, Sleep apnea in acute cerebrovascular diseases: Final report on 128 patients, *Sleep* 22 (2) (1999) 217–223.
74. J. Bamford, P. Sandercock, M. Dennis, C. Warlow, J. Burn, Classification and natural history of clinically identifiable subtypes of cerebral infarction, *The Lancet* 337 (8756) (1991) 1521–1526.

75. M. El-Feky, Radiopaedia, see https://radiopaedia.org/cases/cerebral-venous-infarction-3?lang=us (2019).
76. B. Di Muzio, Radiopaedia, see https://radiopaedia.org/cases/cerebral-venous-haemorrhagic-infarction-1?lang=us (2018).
77. B. Di Muzio, Radiopaedia, see https://radiopaedia.org/cases/cerebral-venous-haemorrhagic-infarction?lang=us (2015).
78. A. Fatemi-Ardekani, C. Boylan, M. D. Noseworthy, Identification of breast calcification using magnetic resonance imaging, *Medical Physics* 36 (12) (2009) 5429–5436.
79. E. A. Sickles, Breast calcifications: Mammographic evaluation, *Radiology* 160 (2) (1986) 289–293.
80. G. Kruger, Radiopaedia, see https://radiopaedia.org/cases/liponecrotic-breast-calcifications?lang=us (2018).
81. G. Kruger, Radiopaedia, see https://radiopaedia.org/cases/dcis-left-breast-10?lang=us (2019).
82. G. Kruger, Radiopaedia, see https://radiopaedia.org/cases/development-of-dcis-right-breast?lang=us (2016).

2

WHO Child Growth Standards Modelling by Variable-, Fractional-Order Difference Equation

Piotr Ostalczyk

CONTENTS

2.1 Introduction 56
2.2 Mathematical Preliminaries 56
    2.2.1 Oblivion Function 57
    2.2.2 Variable-, Fractional-Order Backward Difference 58
    2.2.3 Variable-, Fractional-Order Backward Sum 58
    2.2.4 Variable-, Fractional-Order Backward Difference/Sum Selected Properties 59
        2.2.4.1 Linearity 59
        2.2.4.2 Commutative Law of VFODs and VFOSs 59
        2.2.4.3 Concatenation of VFODs and VFOSs 59
2.3 Variable-, Fractional-Order Dynamical Elements 63
    2.3.1 Variable-, Fractional-Order Discrete Differentiator 63
    2.3.2 Variable-, Fractional-Order Discrete Integrator (Summator) 65
    2.3.3 Variable-, Fractional-Order Discrete Inertial Element 66
    2.3.4 Variable-, Fractional-Order Discrete Summator with Inertia ... 67
2.4 Linear, Time-Invariant Variable-, Fractional-Order Difference Equation 67
2.5 Spline Order Functions 69
    2.5.1 Special Cases 71
        2.5.1.1 First Numerical Example 71
        2.5.1.2 Second Numerical Example 72
2.6 Growth Process Model Parameter Evaluation Problems 76
    2.6.1 Initial Conditions (Birth Parameters) 76
    2.6.2 Nutrition (Nutrition Function) 76
2.7 WHO Population Growth Data 77
2.8 Population Growth Modelling 77
2.9 Conclusions 81
References 81

2.1 Introduction

In this chapter, a discrete fractional calculus is applied to modelling a broadly understood growth of living organisms. Here, one can mention: a population height and weight growth represented by percentile WHO data (WHO Weight-for-age; Castellano and Ho, 2015; Guarino et al., 2015; Kurtz, 2018), tumour growth (Ferrara, 2004; Grimm et al., 2016), plant growth (Fourcaud, 2008), etc.

It is very important nowadays to have a fitting mathematical model accurately describing transients of a real system, especially in the presence of external disturbances and noise. This contemporary 'divination' of computer simulations applied to the models predicts dangers or optimises effects. Methodology based on integer-order differential or difference linear or non-linear equations (Ljung, 1999; Koenker and Hallock, 2001) has been known for years and is still in progress.

In this chapter, a generalisation of the mentioned equations to the variable fractional – order difference linear time – invariant equations is applied to the growth modelling. A short mathematical background concerning the variable-, Grünwald-Letnikov fractional-order backward – difference (Ostalczyk, 2016; Sierociuk and Malesza, 2017) and linear fractional – order difference equation used in the discrete-time dynamical systems description (Caponett et al., 2010; Valerio and da Costa, 2011; Ostalczyk and Mozyrska, 2016a, 2016b; Deb and Roychoudhury, 2019) is given. This type of description is particularly suitable for systems with memory. These types of systems are considered in this chapter. A special matrix-vector form of mentioned equations is used to model the population growth chosen. As the order functions, the spline ones are chosen. Such a choice of the order function makes the equation change its type due to the order-function values. Simplification assumptions are discussed. First, the equation describes the variable-, fractional-order integrator, then the variable-, fractional-order integrator with inertia and finally only the inertial element. The initial conditions are taken into account, and they model the data parameters of a just-born child. Variability of orders enables a simplification of the equation forms and lowers its order, leading to better fitness of the measured to simulated data.

2.2 Mathematical Preliminaries

In this section, some fundamental notions that will be applied in the model are introduced.

2.2.1 Oblivion Function

For an order function $v(l)$ satisfying the condition

$$0 \leq v(l) \leq 1 \text{ for } l \in \mathbb{Z}^+ \cup \{0\} \tag{2.1}$$

one defines a discrete two-variable function $a^{v(l)}(k)$.

Definition 2.1

For a given order function $v(l)$ where $l \in \mathbb{Z}^+ \cup \{0\}$ an oblivion ('decay') function $a^{v(l)}(k)$ is defined by the following formula (Ostalczyk, 2016):

$$a^{v(l)}(k) = \begin{cases} 0 & \text{for } k < 0 \\ 1 & \text{for } k = 0 \\ (-1)^k \dfrac{v(l)[v(l)-1]\cdots[v(l)-k+1]}{k!} & \text{for } k \in \mathbb{Z}^+ \end{cases} \tag{2.2}$$

It can be easily proved that this function's consecutive values can be calculated recursively according to the rule

$$a^{v(l)}(k) = a^{v(l)}(k-1)\left(1 - \frac{v(l)+1}{k}\right) \text{ for } k, l \in \mathbb{Z}^+ \cup \{0\} \tag{2.3}$$

For the order function satisfying the condition

$$-1 \leq -v(l) \leq 0 \text{ for } l \in \mathbb{Z}^+ \cup \{0\} \tag{2.4}$$

The variable-, fractional-order backward sum is defined below.

Definition 2.2

For a given order function $-v(l)$ where $l \in \mathbb{Z}^+ \cup \{0\}$ an oblivion ('decay') function $a^{-v(l)}(k)$ is defined by

$$s^{-v(l)}(k) = \begin{cases} 0 & \text{for } k < 0 \\ 1 & \text{for } k = 0 \\ \dfrac{v(l)[v(l)+1]\cdots[v(l)+k-1]}{k!} & \text{for } k \in \mathbb{Z}^+ \end{cases} \tag{2.5}$$

It can be easily proved that this function's consecutive values can be calculated recursively according to the rule

$$s^{-v(l)}(k) = s^{-v(l)}(k-1)\left(1 + \frac{v(l)-1}{k}\right) \text{ for } k, l \in \mathbb{Z}^+ \cup \{0\} \quad (2.6)$$

The function $s^{-v(l)}(k)$ plays the role of a kernel in the variable-, fractional-order backward difference, whereas $s^{v(l)}(k)$ is the variable-, fractional-order backward sum. They are defined below.

2.2.2 Variable-, Fractional-Order Backward Difference

Definition 2.3

For a given order function satisfying equation (2.4) and related oblivion function (2.5), a variable-, fractional-order Grünwald-Letnikov backward difference (VFOBS) of the bounded discrete variable function $f(k)$ is defined as a sum:

$${}^{GL}_{k_0}\Delta_k^{v(k)} f(k) = \sum_{i=0}^{k-k_0} a^{v(k)}(i) f(k-i)$$

$$= \begin{bmatrix} 1 & a^{v(k)}(1) & \cdots & a^{v(k)}(k-k_0) \end{bmatrix} \begin{bmatrix} f(k) \\ f(k-1) \\ \vdots \\ f(k_0) \end{bmatrix} \quad (2.7)$$

where in VFOBD superscripts GL and $v(k)$ denote the type of backward difference and function order, respectively. The subscripts k_0 and k denote the VFOBD evaluation interval $[k_0, k]$.

2.2.3 Variable-, Fractional-Order Backward Sum

The variable-, fractional-order backward sum (VFOBS) is defined in a similar way.

Definition 2.4

For a given order function satisfying equation (2.1) and related oblivion function (2.2), a VFOBS of the bounded discrete-variable function $f(k)$ is defined as follows:

$$ {}^{GL}_{k_0}\Delta_k^{-v(k)} f(k) = \sum_{i=0}^{k-k_0} a^{-v(k)}(i) f(k-i)$$

WHO Child Growth Standards Modelling

$$= \begin{bmatrix} 1 & a^{-v(k)}(1) & \cdots & a^{-v(k)}(k-k_0) \end{bmatrix} \begin{bmatrix} f(k) \\ f(k-1) \\ \vdots \\ f(k_0) \end{bmatrix} \quad (2.8)$$

2.2.4 Variable-, Fractional-Order Backward Difference/Sum Selected Properties

Here, we prove some useful properties of the VFOBD and VFOBS.

2.2.4.1 Linearity

The VFOBD/VFOBS is a linear operator.

Proposition 2.1

For $a \in \mathbb{R}$ and the discrete function $f(k)$ the VFOD/S satisfies the property

$${}_{k_0}^{GL}\Delta_k^{\pm v(k)} af(k) = a\, {}_{k_0}^{GL}\Delta_k^{\pm v(k)} f(k) \quad (2.9)$$

Proof

From the definition formulae (2.7) and (2.8) one gets the result.

2.2.4.2 Commutative Law of VFODs and VFOSs

In general, for $0 \le v(k), \mu(k) \le 1$ the VFOD/S are not commutative, i.e.,

$$ {}_{k_0}^{GL}\Delta_k^{v(k)} \left[{}_{k_0}^{GL}\Delta_k^{\mu(k)} f(k) \right] \ne {}_{k_0}^{GL}\Delta_k^{\mu(k)} \left[{}_{k_0}^{GL}\Delta_k^{v(k)} f(k) \right] \quad (2.10)$$

2.2.4.3 Concatenation of VFODs and VFOSs

In general, for $0 \le v(k), \mu(k) \le 1$ there is

$$ {}_{k_0}^{GL}\Delta_k^{v(k)} \left[{}_{k_0}^{GL}\Delta_k^{\mu(k)} f(k) \right] \ne {}_{k_0}^{GL}\Delta_k^{v(k)+\mu(k)} f(k) \quad (2.11)$$

The equality is satisfied for $0 \leq v(k) = v = \text{const} \leq 1$ and $0 \leq \mu(k) \leq 1$. This states the following proposition.

Proposition 2.2

For $0 < v(k) = v = \text{const} \leq 1$ and $0 < \mu(k) \leq 1$ the VFOD/S of the discrete function $f(k)$ satisfies the property

$$\underset{k_0}{^{GL}}\Delta_k^v\left[\underset{k_0}{^{GL}}\Delta_k^{\mu(k)} f(k)\right] = \underset{k_0}{^{GL}}\Delta_k^{v+\mu(k)} f(k) \tag{2.12}$$

Proof

For simplicity, it will be assumed that $k_0 = 0$. Such an assumption does not reduce the generality of considerations. The left-hand side of equation (2.12) can be expressed in the form

$$\underset{0}{^{GL}}\Delta_k^v\left[\underset{0}{^{GL}}\Delta_k^{\mu(k)} f(k)\right] = \begin{bmatrix} 1 & a^v(1) & \cdots & a^v(k) \end{bmatrix} \begin{bmatrix} \underset{0}{^{GL}}\Delta_k^{\mu(k)} f(k) \\ \underset{0}{^{GL}}\Delta_{k-1}^{\mu(k)} f(k-1) \\ \vdots \\ \underset{0}{^{GL}}\Delta_0^{\mu(k)} f(0) \end{bmatrix}$$

$$= \begin{bmatrix} 1 & a^v(1) & \cdots & a^v(k) \end{bmatrix} \begin{bmatrix} 1 & a^{\mu(k)}(1) & \cdots & a^{\mu(k)}(k) \\ 0 & 1 & \cdots & a^{\mu(k-1)}(k-1) \\ \vdots & \vdots & & \vdots \\ 0 & 0 & \cdots & 1 \end{bmatrix} \begin{bmatrix} f(k) \\ f(k-1) \\ \vdots \\ f(0) \end{bmatrix} \tag{2.13}$$

To prove equation (2.12), one should prove that

$$\begin{bmatrix} 1 & a^v(1) & \cdots & a^v(k) \end{bmatrix} \begin{bmatrix} a^{\mu(k)}(k) \\ a^{\mu(k)}(k-1) \\ \vdots \\ 1 \end{bmatrix} = a^{v+\mu(k)}(k) \tag{2.14}$$

WHO Child Growth Standards Modelling 61

is satisfied for $k, k-1, \cdots, 0$, so there is $k+1$ equations of the type (2.14). Collecting all the mentioned equations in one vector-matrix form one gets

$$\begin{bmatrix} 1 & a^v(1) & \cdots & a^v(k) \\ 0 & 1 & \cdots & a^v(k-1) \\ \vdots & \vdots & & \vdots \\ 0 & 0 & \cdots & 1 \end{bmatrix} \begin{bmatrix} a^{\mu(k)}(k) \\ a^{\mu(k)}(k-1) \\ \vdots \\ 1 \end{bmatrix} = \begin{bmatrix} a^{v+\mu(k)}(k) \\ a^{v+\mu(k)}(k-1) \\ \vdots \\ 1 \end{bmatrix} \quad (2.15)$$

Now, according to equation (2.3) one has

$$a^v(k) = a^v(k-1)\left(1 - \frac{v+1}{k}\right) \quad (2.16)$$

$$d^{v+\mu(k)}(k) = d^{v+\mu(k)}(k-1)\left(1 - \frac{v+\mu(k)+1}{k}\right) \quad (2.17)$$

so

$$\begin{bmatrix} 1 & a^v(1) & \cdots & a^v(k-1)\left(1-\frac{v+1}{k}\right) \\ 0 & 1 & \cdots & a^v(k-1) \\ \vdots & \vdots & & \vdots \\ 0 & 0 & \cdots & 1 \end{bmatrix} \begin{bmatrix} a^{\mu(k)}(k) \\ a^{\mu(k)}(k-1) \\ \vdots \\ 1 \end{bmatrix}$$

$$= \begin{bmatrix} a^{v+\mu(k)}(k-1)\left(1-\frac{v+\mu(k)+1}{k}\right) \\ a^{v+\mu(k)}(k-1) \\ \vdots \\ 1 \end{bmatrix} \quad (2.18)$$

Now, performing elementary operations on the first row of the above expression to the first row we add the second one multiplied by $\frac{v+\mu(k)+1}{k} - 1$. The result

$$\begin{bmatrix} 1 & \dfrac{(v+1)(1-k)+\mu(k)}{k} & \cdots & a^v(k-1)\dfrac{\mu(k)}{k} \\ 0 & 1 & \cdots & d^v(k-1) \\ \vdots & \vdots & & \vdots \\ 0 & 0 & \cdots & 1 \end{bmatrix} \begin{bmatrix} a^{\mu(k)}(k) \\ a^{\mu(k)}(k-1) \\ \vdots \\ 1 \end{bmatrix}$$

$$= \begin{bmatrix} 0 \\ a^{v+\mu(k)}(k-1) \\ \vdots \\ 1 \end{bmatrix} \qquad (2.19)$$

is essentially simplified with respect to equation (2.18). Now, by simple but tedious calculations one confirms the equality

$$a^{\mu(k)}(k) + \dfrac{(v+1)(1-k)+\mu(k)}{k} a^{\mu(k)}(k-1) + \cdots + a^v(k-1)\dfrac{\mu(k)}{k} = 0 \quad (2.20)$$

Remark 2.1

Unfortunately,

$${}^{GL}_{k_0}\Delta_k^{\mu(k)}\left[{}^{GL}_{k_0}\Delta_k^v f(k)\right] \neq {}^{GL}_{k_0}\Delta_k^{\mu(k)+v} f(k) \qquad (2.21)$$

Remark 2.2

Similar conditions are satisfied for $-1 \leq -v(k) = -v = \text{const} < 0$ and $-1 \leq -\mu(k) < 0$.

Remark 2.3

As a consequence of (2.11) there is

$${}^{GL}_{k_0}\Delta_k^{v(k)}\left[{}^{GL}_{k_0}\Delta_k^{-v(k)} f(k)\right] \neq {}^{GL}_{k_0}\Delta_k^{-v(k)}\left[{}^{GL}_{k_0}\Delta_k^{v(k)} f(k)\right] \neq f(k) \qquad (2.22)$$

2.3 Variable-, Fractional-Order Dynamical Elements

Now, fundamental variable-, fractional-order linear time invariant elements will be recalled. In (Mozyrska and Ostalczyk, 2016a, 2016b; Ostalczyk, 2016a, 2016b; Ostalczyk and Mozyrska, 2017) there were discussed: the variable-, fractional-order discrete time integrator oscillation and inertial element, respectively. A novelty integrator with inertia will be defined. In all elements, the output signal will be denoted as $y(kh)$, whereas the input signal will be denoted as $u(kh)$. Here, h denotes the sampling period. When h is omitted it will be assumed that $h = 1$. The following nomenclature will be used: **V** – variable, **F** – fractional, **I** – integer, **O** – order, **d** – differentiator, **i** – integrator, **i** – inertia.

2.3.1 Variable-, Fractional-Order Discrete Differentiator

The variable-, fractional-order discrete differentiator (VFOD) is described by a variable-, fractional-order difference equation

$$ {}^{GL}_{k_0}\Delta_k^{-v(k)} y(kh) = b_0 u(kh) \qquad (2.23) $$

where $-1 \leq -v(k) < 0$ and b_0 is a constant coefficient. The preceding equation can be expressed in a vector form

$$ \begin{bmatrix} 1 & a^{-v(k)}(1) & \cdots & a^{-v(k)}(k-k_0) \end{bmatrix} \begin{bmatrix} y(kh) \\ y(kh-h) \\ \vdots \\ y(k_0 h) \end{bmatrix} = b_0 u(kh) \qquad (2.24) $$

and is valid for all $k, k-1, \cdots, 1, 0$. Collecting such equations in one vector-matrix form one obtains

$$\begin{bmatrix} 1 & a^{-v(k)}(1) & \cdots & a^{-v(k)}(k-k_0-1) & a^{-v(k)}(k-k_0) \\ 0 & 1 & \cdots & a^{-v(k-1)}(k-k_0-2) & a^{-v(k-1)}(k-k_0-1) \\ \vdots & \vdots & & \vdots & \vdots \\ 0 & 0 & \cdots & 1 & a^{-v(1)}(1) \\ 0 & 0 & \cdots & 0 & 1 \end{bmatrix} \begin{bmatrix} y(kh) \\ y(kh-h) \\ \vdots \\ y(k_0h) \end{bmatrix}$$

$$= b_0 \begin{bmatrix} u(kh) \\ u(kh-h) \\ \vdots \\ u(k_0h+h) \\ u(k_0h) \end{bmatrix}$$
(2.25)

For simplicity, the preceding equation will be denoted as

$$_{k_0}\mathbf{D}_k^{-v(k)} y(k) = b_0 u(k) \qquad (2.26)$$

where

$$_{k_0}\mathbf{D}_k^{-v(k)} = \begin{bmatrix} 1 & a^{-v(k)}(1) & \cdots & a^{-v(k)}(k-1) & a^{-v(k)}(k) \\ 0 & 1 & \cdots & a^{-v(k-1)}(k-2) & a^{-v(k-1)}(k-1) \\ \vdots & \vdots & & \vdots & \vdots \\ 0 & 0 & \cdots & 1 & a^{-v(1)}(1) \\ 0 & 0 & \cdots & 0 & 1 \end{bmatrix}, \qquad (2.27)$$

$$y(k) = \begin{bmatrix} y(kh) \\ y(kh-h) \\ \vdots \\ y(k_0h) \end{bmatrix}, \; u(k) \begin{bmatrix} u(kh) \\ u(kh-h) \\ \vdots \\ u(k_0h+h) \\ u(k_0h) \end{bmatrix} \qquad (2.28)$$

Square $(k+1) \times (k+1)$ matrix $_{k_0}\mathbf{D}_k^{-v(k)}$ is always non-singular, so equation (2.26) can be expressed in an equivalent form

$$y(kh) = b_0 \left[_{k_0}\mathbf{D}_k^{-v(k)} \right]^{-1} u(kh) \qquad (2.29)$$

Using vector-matrix notation one defines the second form of the VFOd

$$y(kh) = b_0 \,_{k_0}\mathbf{D}_k^{v(k)} u(kh) \tag{2.30}$$

Here, one should recall, that according to formula (2.22) $[_{k_0}\mathbf{D}_k^{-v(k)}]^{-1} \neq \,_{k_0}\mathbf{D}_k^{v(k)}$. Finally, one can define the third form of the variable-, fractional-order discrete differentiator for $0 < v_1(k), v_2(k) \leq 1$

$$_{k_0}\mathbf{D}_k^{-v_1(k)} y(kh) = b_0 \mathbf{D}_k^{v_2(k)} u(kh) \tag{2.31}$$

which can be always expressed in the form

$$y(kh) = b_0 \left[_{k_0}\mathbf{D}_k^{-v_1(k)} \right]^{-1} \mathbf{D}_k^{v_2(k)} u(kh) \tag{2.32}$$

2.3.2 Variable-, Fractional-Order Discrete Integrator (Summator)

To distinguish the variable-, fractional-order discrete integrator (VFOi) from the VFOD, the order function of the first mentioned will be denoted as $\mu(k)$. It satisfies a condition $0 < \mu(k) \leq 1$. There are also three types of VFOi's (Ostalczyk, 2016). They are described by the following equations:

$$_{k_0}\mathbf{D}_k^{\mu(k)} y(kh) = b_0 u(kh) \tag{2.33}$$

$$y(kh) = b_0 \,_{k_0}\mathbf{D}_k^{-\mu(k)} u(kh) \tag{2.34}$$

$$_{k_0}\mathbf{D}_k^{\mu_1(k)} y(k) = b_0 \,_{k_0}\mathbf{D}_k^{-\mu_2(k)} u(k) \tag{2.35}$$

for $0 < \mu_1(k), \mu_2(k) \leq 1$. As in the VFOD case, all matrices in equations (2.33) and (2.35) are invertible. This property enables to transform equations (2.33) and (2.35) to the equivalent forms

$$y(k) = b_0 \left[_{k_0}\mathbf{D}_k^{\mu(k)} \right]^{-1} u(k) \tag{2.36}$$

$$y(k) = b_0 \left[_{k_0}\mathbf{D}_k^{\mu_1(k)} \right]^{-1} {}_{k_0}\mathbf{D}_k^{-\mu_2(k)} u(k) \tag{2.37}$$

Remark 2.4
Here, one should realise that in general there is no order function $0 < \xi(k) \le 2$ for which $\mathbf{D}_k^{\xi(k)} = [\mathbf{D}_k^{-\mu_2(k)}]^{-1}\mathbf{D}_k^{\mu_1(k)}$ for $0 < \mu_1(k), \mu_2(k) \le 1$.

2.3.3 Variable-, Fractional-Order Discrete Inertial Element

Following the differential equation describing the first (integer)-order inertial element (IOi), a similar one defines a variable-, fractional-order inertial (VFOi) element using VFOBD. Its equation is given below:

$$\substack{GL\\k_0}\Delta_k^{v(k)}y(k) + a_0 y(k) = b_0 u(k) \tag{2.38}$$

where a_0, b_0 are constant parameters. Following the proposed vector-matrix notation equivalent description of the considered element is as follows:

$$\substack{k_0}\mathbf{D}_k^{v(k)}y(k) + a_0 y(k) = b_0 u(k) \tag{2.39}$$

Here, a VFOi can be generalised for $0 < v_1(k), v_2(k) \le 1$

$$\substack{k_0}\mathbf{D}_k^{v_1(k)}\left[\substack{k_0}\mathbf{D}_k^{-v_2(k)}\right]^{-1}y(k) + a_0 y(k) = b_0 u(k) \tag{2.40}$$

or

$$\left[\substack{k_0}\mathbf{D}_k^{-v_2(k)}\right]^{-1}\substack{k_0}\mathbf{D}_k^{v_1(k)}y(k) + a_0 y(k) = b_0 u(k) \tag{2.41}$$

For VFOi, one can impose the additional condition

$$0 < \mu(k) + v(k) \le 1 \tag{2.42}$$

Equation (2.40) can be expressed in the form

$$\substack{k_0}\mathbf{D}_k^{v(k)}\left[\substack{k_0}\mathbf{D}_k^{-\mu(k)}\right]^{-1}y(k) + a_0 1_k y(k) = b_0 u(k) \tag{2.43}$$

where 1_k is $(k+1) \times (k+1)$ identity matrix. Hence, it can be transformed to an equivalent form

$$\left\{\substack{k_0}\mathbf{D}_k^{v(k)}\left[\substack{k_0}\mathbf{D}_k^{-\mu(k)}\right]^{-1} + a_0 1_k\right\}y(k) = b_0 u(k) \tag{2.44}$$

Matrices ${}_{ko}\mathbf{D}_k^{v(k)}$ and ${}_{ko}\mathbf{D}_k^{-\mu(k)}$ are upper triangular with units on the main diagonal, so assuming that

$$1 + a_0 \neq 0 \tag{2.45}$$

Equation (2.44) gets the form

$$y(k) = b_0 \left\{ {}_{ko}\mathbf{D}_k^{v(k)} \left[{}_{ko}\mathbf{D}_k^{-\mu(k)} \right]^{-1} + a_0 \mathbf{1}_k \right\}^{-1} u(k) \tag{2.46}$$

This is a solution for the Variable-, Fractional-Order Difference Equation (VFODE) with zero initial conditions to the known bounded input function (k).

2.3.4 Variable-, Fractional-Order Discrete Summator with Inertia

Finally, a variable-, fractional-order discrete integrator (summator) with inertia (VFOii) is defined. For $0 < v_1(k) + v_2(k) \leq 1$ it is described by the equation

$${}_{ko}\mathbf{D}_k^{v_1(k)} \left[{}_{ko}\mathbf{D}_k^{v_2(k)} y(k) + a_0 y(k) \right] = {}_{ko}\mathbf{D}_k^{v_1(k)} {}_{ko}\mathbf{D}_k^{v_2(k)} y(k) + a_0 {}_{ko}\mathbf{D}_k^{v_1(k)} y(k) = b_0 u(k) \tag{2.47}$$

Assuming that condition (2.45) is satisfied, one can find a solution

$$\begin{aligned} y(k) &= b_0 \left\{ {}_{ko}\mathbf{D}_k^{v_1(k)} \left[{}_{ko}\mathbf{D}_k^{v_2(k)} y(k) + \mathbf{1}_k a_0 \right] \right\}^{-1} u(k) \\ &= \left[{}_{ko}\mathbf{D}_k^{v_2(k)} y(k) + \mathbf{1}_k a_0 \right]^{-1} \left[{}_{ko}\mathbf{D}_k^{v_1(k)} \right]^{-1} u(k) \end{aligned} \tag{2.48}$$

2.4 Linear, Time-Invariant Variable-, Fractional-Order Difference Equation

Considered in Chapter 2, dynamical elements are the special cases of the general dynamical element. This is described by the linear, time-invariant variable-, fractional-order difference equation

$$\sum_{i=0}^{n} a_i \, {}_{ko}\Delta_k^{v_i(k)} y(k) = \sum_{j=0}^{m} b_j \, {}_{ko}\Delta_k^{\mu_i(k)} u(k) \tag{2.49}$$

or in the vector-matrix convention

$$\sum_{i=0}^{n} a_{i\,k_0} \mathbf{D}_k^{v_i(k)} y(k) = \sum_{j=0}^{m} b_{j\,k_0} \mathbf{N}_k^{\mu_i(k)} u(k) \qquad (2.50)$$

where

$$a_n = 1$$

$$1 + \sum_{i=0}^{n-1} a_i \neq 0 \qquad (2.51)$$

and the order functions are arranged as in the common difference equations

$$v_n(k) > v_{n-1}(k) > \cdots > v_1(k) > v_0(k) = 0$$

$$\mu_m(k) > \mu_{m-1}(k) > \cdots > \mu_1(k) > \mu_0(k)$$

$$n \geq m \qquad (2.52)$$

It is assumed that the input vector $u(k)$ is known and $u(k) = 0$ for $k < k_0$. In the general solution of equation (2.50), initial conditions must be taken into account (Sierociuk et al., 2015). They are collected in $\infty \times 1$ initial conditions vector

$$y_{k_0-1}(k) = \begin{bmatrix} y_{k_0-1} \\ y_{k_0-2} \\ y_{k_0-3} \\ \vdots \end{bmatrix} \qquad (2.53)$$

Then, the general solution is of the form

$$\left[\sum_{i=0}^{n} a_{i\,k_0} \mathbf{D}_k^{v_i(k)} \quad \sum_{i=0}^{n} a_{i\,-\infty} \mathbf{D}_{k_0-1}^{v_i(k)} \right] \begin{bmatrix} y(k) \\ y_{k_0-1}(k) \end{bmatrix} = \sum_{j=0}^{m} b_j \mathbf{N}_k^{\mu_i(k)} u(k) \qquad (2.54)$$

which is further rearranged as

$$\sum_{i=0}^{n} a_{i\,k_0} \mathbf{D}_k^{v_i(k)} y(k) = \sum_{j=0}^{m} b_j \mathbf{N}_k^{\mu_i(k)} u(k) - \sum_{i=0}^{n} a_{i\,-\infty} \mathbf{D}_{k_0-1}^{v_i(k)} y_{k_0-1}(k) \qquad (2.55)$$

Finally, the solution is

$$\sum_{i=0}^{n} a_{i\,k_0} \mathbf{D}_k^{v_i(k)} \mathbf{y}(k) = \left[\sum_{i=0}^{n} a_{i\,k_0} \mathbf{D}_k^{v_i(k)}\right]^{-1} \sum_{j=0}^{m} b_j \mathbf{N}_k^{\mu_i(k)} \mathbf{u}(k) \qquad (2.56)$$
$$- \left[\sum_{i=0}^{n} a_{i\,k_0} \mathbf{D}_k^{v_i(k)}\right]^{-1} \sum_{i=0}^{n} a_{i\,-\infty} \mathbf{D}_{k_0-1}^{v_i(k)} \mathbf{y}_{k_0-1}(k)$$

Remark 2.5

The initial condition $(k - k_0 + 1) \times \infty$ matrices for $i = 1, 2, \cdots, n$ are of the form

$$_{-\infty}\mathbf{D}_{k_0-1}^{v_i(k)} = \begin{bmatrix} a^{v_i(k)}(k-k_0+1) & a^{v_i(k)}(k-k_0+2) & a^{v_i(k)}(k-k_0+3) & \cdots \\ a^{v_i(k-1)}(k-k_0) & a^{v_i(k-1)}(k-k_0+1) & a^{v_i(k-1)}(k-k_0+2) & \cdots \\ \vdots & \vdots & \vdots & \cdots \\ a^{v_i(1)}(2) & a^{v_i(1)}(3) & a^{v_i(1)}(4) & \cdots \\ a^{v_i(0)}(1) & a^{v_i(0)}(2) & a^{v_i(0)}(3) & \cdots \end{bmatrix}$$
(2.57)

Their elements rapidly tend to 0, that is, $a^{v_i(j)}(l) \to 0$ for $l \to +\infty$. This means that in practice only a few bounded initial conditions can be taken into account.

2.5 Spline Order Functions

Conditions (2.52) were imposed on the considered order functions. These may be expressed in another form

$$\prod_{i=1}^{n}\left(_{k_0}\Delta_k^{v_i(k)} + a_i'\right) y(k) = \prod_{j=1}^{m}\left(_{k_0}\Delta_k^{\mu_j(k)} + b_j'\right) u(k) \qquad (2.58)$$

or equivalently

$$\prod_{i=1}^{n}\left(_{k_0}\mathbf{D}_k^{v_i(k)} + 1_k a_i'\right) y(k) = \prod_{j=1}^{m}\left(_{k_0}\mathbf{N}_k^{\mu_j(k)} + 1_k b_j'\right) u(k) \qquad (2.59)$$

with conditions

$$0 \le v_i(k) \le 1 \text{ for } i = 1, 2, \cdots, n$$

$$-1 \le \mu_j(k) \le 1 \text{ for } j = 1, 2, \cdots, m \tag{2.60}$$

Remark 2.6

For $-1 \le \mu_j(k) \le 0$ the condition $m > n$ is admissible.

Remark 2.7

For

$$v_i(k) = 1 \text{ for } i = 1, 2, \cdots, n$$

$$\mu_j(k) = 1 \text{ for } j = 1, 2, \cdots, m \tag{2.61}$$

Equation (2.59) is a classical Integer order (IO) difference equation. In matrix (2.57), all elements $a^{v_i(k)}(n+l) = 0$ for $l = 1, 2, \cdots$, so only n initial conditions are taken into account.

Similarly, one can express a solution

$$y(k) = \left[\prod_{i=1}^{n} \left({}_{k_0}\mathbf{D}_k^{v_i(k)} + 1_k a_i' \right) \right]^{-1} \prod_{j=1}^{m} \left({}_{k_0}\mathbf{N}_k^{\mu_j(k)} + 1_k b_j' \right) u(k) \tag{2.62}$$

or taking into account the rules of matrix inversion

$$y(k) = \prod_{i=n}^{1} \left({}_{k_0}\mathbf{D}_k^{v_i(k)} + 1_k a_i' \right)^{-1} \prod_{j=1}^{m} \left({}_{k_0}\mathbf{N}_k^{\mu_j(k)} + 1_k b_j' \right) u(k) \tag{2.63}$$

Evidently, there are relations between the order functions and coefficients. The simplest ones are given below:

$$v_n(k) = \prod_{i=1}^{n} v_i'(k) \tag{2.64}$$

WHO Child Growth Standards Modelling

$$a_0 = \prod_{i=1}^{n} a'_i \tag{2.65}$$

It is convenient to define an order function as a spline one

$$v'_i(k) = \begin{cases} v'_{i1}(k) & \text{for} & k_0 \leq k < k_1 \\ v'_{i2}(k) & \text{for} & k_1 \leq k < k_2 \\ \vdots & & \vdots \\ v'_{i,w-1}(k) & \text{for} & k_{w-2} \leq k < k_{w-1} \\ v'_{iw}(k) & \text{for} & k > k_{w-1} \end{cases} \tag{2.66}$$

Remark 2.8

To preserve the 'smoothness' of the solution, it is recommended to add conditions

$$v'_{i,j-1}(k_{j-1}-1) = v'_{i,j}(k_{j-1}) \text{ for } j = 2, \cdots, w \tag{2.67}$$

2.5.1 Special Cases

One considers a special case of equation (2.59) with $n = 2$, $m = 1$, $a'_1 \neq 0, a'_2 = 0$, $b'_1 \neq 0$, $\mu'_1(k) = 0$. Then, the VFODE has the form

$$_{k_0}\mathbf{D}_k^{v'_2(k)}\left(_{k_0}\mathbf{D}_k^{v'_1(k)} + 1_k a'_1\right)y(k) = \left(1 + b'_1\right)u(k) \tag{2.68}$$

and according to combinations of the order functions $v'_1(k), v'_2(k)$, the VFODE elements can pupate to various intermediate forms (Mozyrska and Ostalczyk, 2016; Ostalczyk and Mozyrska, 2016a, 2016b, 2017). These are mentioned in Table 2.1.

In the following two numerical examples, some aspects of the VFODE are presented.

2.5.1.1 First Numerical Example

One considers VFOi (2.59) with a spline order function (position 2 in Table 2.1)

$$v'_2(k) = \begin{cases} 0.5 & \text{for} & 0 \leq k < 60 \\ 0.5 - \sin\left(\dfrac{\pi k}{12}\right) & \text{for} & 60 \leq k < 119 \\ 1 & \text{for} & k \geq 119 \end{cases} \tag{2.69}$$

TABLE 2.1
VFODE and Its Special Forms

No.	Order Functions for $k \in \mathbb{Z}^+ \cup \{0\}$	Type of Element
1	$v_1'(k) = 0$ $v_2'(k) = 1$	IOi
2	$v_1'(k) = 0$ $0 < v_2'(k) \leq 1$	VFOi
3	$v_1'(k) = 1$ $v_2'(k) = 0$	IOi
4	$0 < v_1'(k) \leq 1$ $v_2'(k) = 0$	VFOi
5	$v_1'(k) = 1$ $v_2'(k) = 1$	IOii
6	$0 < v_1'(k) \leq 1$ $0 < v_2'(k) \leq 1$	VFOii

The plot of the order function is given in Figure 2.1.

The output signal to the unit step function, zero initial conditions and $b_0 = 1$ is presented in Figure 2.2.

Remark 2.9

In the first-time interval $[0, 60)$ the VFOi is a FOi of an order $v_2'(k) = 0.5$. Next, for $k \in [50, 119)$ there is a sine order function strongly influencing the solution. Finally, for $k \geq 119$ the dynamical system is an IOi (classical first-order discrete integrator).

2.5.1.2 Second Numerical Example

One considers the VFOi (2.68) with spline order function defined for $v_1'(k)$

$$v_1'(k) = \begin{cases} \dfrac{v_1}{k_1^2} k^2 & \text{for} \quad 0 \leq k < k_1 \\ ak^2 + bk + c & \text{for} \quad k_1 \leq k < k_2 \\ v_2 & \text{for} \quad k \geq k_2 \end{cases} \quad (2.70)$$

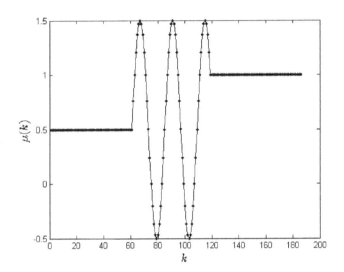

FIGURE 2.1
Plot of the VFOi order function.

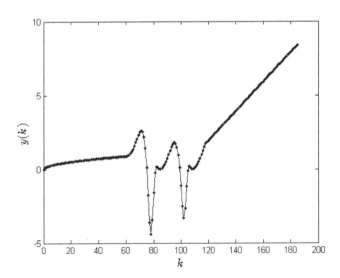

FIGURE 2.2
Plot of the VFOi output signal to the unit step function and zero initial conditions.

and constant parameters k_1, k_2 and v_1, v_2 where

$$\begin{bmatrix} a \\ b \\ c \end{bmatrix} = \begin{bmatrix} 2k_2 & 1 & 0 \\ k_2^2 & k_2 & 1 \\ k_1^2 & k_1 & 1 \end{bmatrix}^{-1} \begin{bmatrix} 0 \\ v_2 \\ v_1 \end{bmatrix} \qquad (2.71)$$

For $v_2 = 2v_1$ and $k_2 = 2k_1$ one has

$$v_i'(k) = \begin{cases} \dfrac{v_1}{k_1^2} k^2 & \text{for} \quad 0 \le k < k_1 \\ ak^2 + bk + c & \text{for} \quad k_1 \le k < 2k_1 \\ 2v_1 & \text{for} \quad k \ge 2k_1 \end{cases} \qquad (2.72)$$

with

$$\begin{bmatrix} a \\ b \\ c \end{bmatrix} = \begin{bmatrix} 4k_1 & 1 & 0 \\ 4k_1^2 & 2k_1 & 1 \\ k_1^2 & k_1 & 1 \end{bmatrix}^{-1} \begin{bmatrix} 0 \\ 2v_1 \\ v_1 \end{bmatrix} = \frac{1}{k_1^2}\begin{bmatrix} k_1 & -1 & 0 \\ -3k_1^2 & 4k_1 & -4k_1 \\ 2k_1^3 & -3k_1^2 & 4k_1^2 \end{bmatrix} \begin{bmatrix} 0 \\ 2v_1 \\ v_1 \end{bmatrix} = \begin{bmatrix} -\dfrac{2v_1}{k_1^2} \\ \dfrac{4v_1}{k_1} \\ -2v_1 \end{bmatrix}$$

(2.73)

For $v_1 = 0.5$, $v_2 = 1.0$, $k_1 = 60$, $k_2 = 120$ the order function and the unit step response are presented in Figures 2.3 and 2.4.

Remark 2.10

In Figures 2.1 through 2.4, the independent variable k denotes any unit of time, day, week, month or year. Here, in the two preceding numerical examples, the variety of shaping transient characteristics options is presented.

WHO Child Growth Standards Modelling

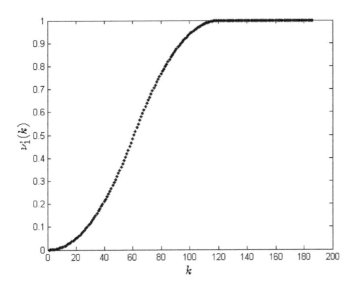

FIGURE 2.3
Plot of the VFOi order function.

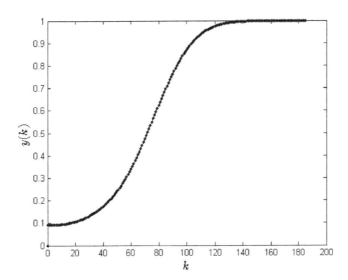

FIGURE 2.4
Plot of the VFOi output signal to the unit step function and zero initial conditions.

2.6 Growth Process Model Parameter Evaluation Problems

It is evident that the growth process model depends on many factors (living conditions, stress, family genetic conditioning, etc.). Hence, the process can be treated as multi-input, multi-output. Here, a single input single-output model will be proposed. In addition, the birth data should be taken into account.

2.6.1 Initial Conditions (Birth Parameters)

It is evident that at the birth moment, a child is of non-zero weight and height. Therefore, these parameters should be treated as the initial conditions in the mathematical model. In the variable-, fractional-order model, there is a possibility to consider not only one value. More than one initial condition should be taken into account during the pregnancy period.

2.6.2 Nutrition (Nutrition Function)

The weight and the height treated as outputs of the 'dynamical system' are subjected to many inputs signals. Here, the most important is nutrition. It may be measured in daily meal weight in $[g]$ or in $[kcal]$ as the function of the age of the child (The Basal Metabolic Rates of Women, 2019). Assuming that $g \propto kcal$ one can model the input as (Figure 2.5)

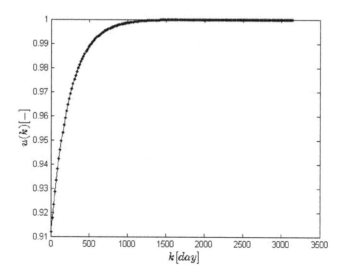

FIGURE 2.5
Plot of the hypothetical nutrition input function $u(k)$.

2.7 WHO Population Growth Data

There are many sources of the population growth data. The most reliable are the data presented by the World Health Organization (WHO). They include percentile charts serving to evaluate the correctness of rates: weight for age, height to age (birth to 6 months, birth to 2 years, birth to 5 years) (WHO, 2019). Every plot of the percentile chart is marked by an integer number from an exemplary set: $\{3, 10, 25, 50, 75, 90, 97\}$. Plots presented in Figure 2.6 are continuous-time functions relating the population growth data.

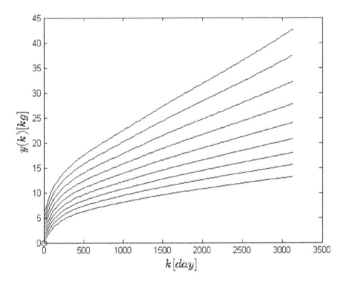

FIGURE 2.6
Weight vs. days.

2.8 Population Growth Modelling

A special case of equation (2.59) is considered with $n = 1$, $m = 2$, $v_1'(k) = v(k) > 0$, $a_1' = 0$, $\mu_2'(k) = 0$, $\mu_1'(k) = \mu(k) < 0$, $b_1' = 0$, $b_2' + 1 = b_0 \neq 0$. Hence, the following model is inferred

$$_{k_0}\mathbf{D}_k^{v(k)} y(k) = b_{0\,k_0} \mathbf{N}_k^{\mu(k)} u(k) \tag{2.74}$$

The proposed model is more versatile in comparison to the fractional (constant)-order and evidently integer-order models. As the order functions, one takes

$$v(k) = \begin{cases} 1.01 + 0.78e^{-0.21(k-1)} & \text{for} \quad 0 \le k < 510 \\ 1 + 0.7e^{-0.1k} & \text{for} \quad 510 \le k \end{cases}$$

$$\mu(k) = \begin{cases} -0.189e^{-0.053(k-1)} & \text{for} \quad 0 \le k < 850 \\ -0.19e^{-0.049k} & \text{for} \quad 850 \le k \end{cases} \quad (2.75)$$

Plots of the order functions are given in Figure 2.7. The simulation result $y_s(k)$ for $b_0 = 0.11$ and $u(k) = 1 - 0.0881e^{-0.07(k-1)}$ (with a solid line) together with the given data $y_d(k)$ (dashed line) are presented in Figure 2.8. There are almost invisible differences between them. Hence, showing those two error functions is defined.

An error defined as the difference between measured (collected) data and simulated ones is taken into account

$$e(k) = y_s(k) - y_d(k) \left[\text{kg} \right] \quad (2.76)$$

It is characterised by its maximal value $\max|e(k)|$. A relative error is also analysed

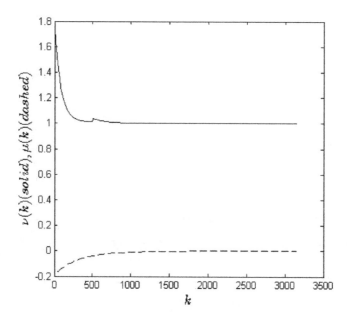

FIGURE 2.7
The order functions for k[day].

WHO Child Growth Standards Modelling

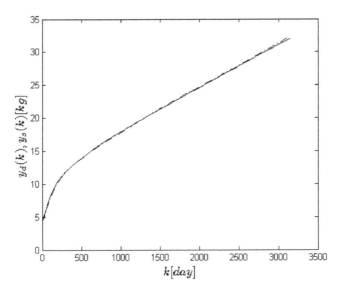

FIGURE 2.8
Simulated $y_s(k)$ (solid line) and given $y_d(k)$ (dotted line) data plots.

$$e_{\text{rel}}(k) = \left[\frac{y_s(k)}{y_d(k)} - 1\right] 100\,[\%] \tag{2.77}$$

The differences between $y_s(k)$ and $y_d(k)$ are almost invisible. The errors (2.76) and (2.77) are plotted in Figures 2.9 and 2.10.

Remark 2.11

It should be noted that although there are still no uniform stability criteria for variable-, fractional-order systems, conditions ensuring stability of such systems may be specified. For order functions satisfying limits

$$\lim_{k \to +\infty} v(k) = \text{const} \in \mathbb{Z}$$

$$v(k) = \text{const} \in \mathbb{Z} \text{ for } k > k_0$$

Such orders were considered in the modelling.

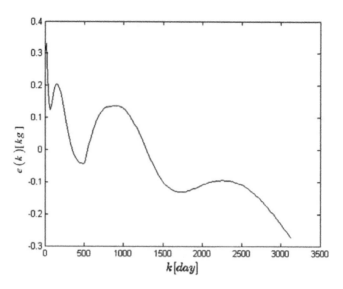

FIGURE 2.9
Error (2.76) plot.

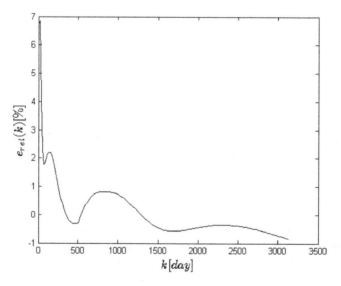

FIGURE 2.10
Relative error (2.77) plot.

2.9 Conclusions

The proposed variable-, fractional-order model was optimised due to the integral of squared error in the discrete version where an integral is replaced by a sum. The optimal model parameters were obtained for the variable-, fractional-order model. One should mention that constant and integer orders were also acceptable. Evidently, an integer order model with a higher number of parameters will be better, but it will always be inferior when volatility and fractionality of orders are allowed.

In this paper, only growth of the dynamic process was modelled. It may serve in the control of it in the closed-loop system. It may give optimal results due to assumed optimality criteria or the presence of external uncontrolled disturbances. This was not a subject considered in this chapter. Having a good mathematical model, one can essentially remove the risks of growth or disease.

Some improvements of the proposed model are mentioned below:

- Additional initial conditions should be added in the model.
- Nutrition function should be precisely specified.
- More inputs should be added leading to the multi-input one output dynamical system.
- In percentile charts, there are non-linearities. Therefore, one should introduce the static or dynamic non-linearity.

References

The Basal metabolic Rates of Women. (2019). https://fitfolk.com/wp content/uploads/2017/02/femaleBMR.png.

Caponetto, R., Dongola, R., Fortuna, G. and Petras, I. (2010). *Fractional Order Systems: Modeling and Control Applications*, World Scientific, Singapore.

Castellano, K.E. and Ho, A.D. (2015). Practical differences among aggregate-level conditional status metrics from median student growth percentiles to value-added models. *Journal of Educational and Behavioral Statistics*, 40, 35–68.

Deb, A. and Roychoudhury, S. (2019). *Control System Analysis and Identification with Matlab®, Block Pulse and Related Orthogonal Functions*, CRC Press, Boca Raton, FL.

Ferrara, N. (2004). Vascular endothelial growth factor basic science and clinical progress. *Endocrine Reviews*, 25(44), 581–611.

Fourcaud, T., Zhang, X., Stokes, A., Lambers, H. and Körner, C. (2008). Plant growth modelling and applications: The increasing importance of plant architecture in growth models. *Annals of Botany*, 101(8), 1053–1065.

Grimm, K.J., Ram, N. and Eswtabrook, R. (2016). *Growth Modeling: Structural Equation and Multilevel Modeling Approaches*, Guilford Press, New York.

Guarino, C., Reckase, M., Stacy, B. and Wooldridge, J. (2015). A comparison of student growth percentile and value-added models of teacher performance. *Statistics and Public Policy*, 2, 1–11.

Koenker, R. and Hallock, K. (2001). Quantile regression: An introduction. *Journal of Economic Perspectives*, 15, 43–56.

Kurtz, M.D. (2018). Value-added and student growth percentile models: What drives differences in estimated classroom effects? *Statistics and Public Policy*, 5(1), 1–8.

Ljung, L. (1999). *System Identification: Theory for the User*, Prentice Hall, Upper Saddle River, NJ.

Mozyrska, D. and Ostalczyk, P. (2016). The second form of the variable-, fractional-order discrete-time integrator. In *Proceedings of the 21st International Conference on Methods and Models in Automation and Robotics (MMAR) IEEE Xplore Digital Library*, INSPEC Accession Number: 16338728, IEEE. doi:10.1109/MMAR.2016.7575250.

Ostalczyk, P. (2016). *Discrete Fractional Calculus: Applications in Control and Image Processing*. World Scientific, Series in Computer Vision–vol. 4, Singapore.

Ostalczyk, P. and Mozyrska, D. (2016a). Variable-fractional-order linear time-invariant system – description and response. In *Proceedings of the International Conference on Fractional Differentiation and Its Applications*, Novi Sad, Serbia, July 18–20, 799–807.

Ostalczyk, P. and Mozyrska, D. (2016b). Variable-, fractional-order oscillation element: Theory and applications of non-integer order systems. In *8th Conference on Non-Integer Order Calculus and Its Applications*, Zakopane, Poland, pp. 65–76.

Ostalczyk, P. and Mozyrska, D. (2017). Variable-, fractional-order inertial element, theory and applications of non-integer order systems. *8th Conference on Non-Integer Order Calculus and Its Applications*, Łódź, Ed. Artur Babiarz, Adam Czornik, Jerzy Klamka, Michał Niezabitowski. Springer Verlag, Lecture Notes in Electrical Engineering, vol. 407, pp. 77–87.

Sierociuk, D. and Malesza, W. (2017). Fractional variable order discrete-time systems, their solutions and properties. *International Journal of System Sciences*, 48(14), 3098–3105. doi:10.10080/00207721.2017.1365969.

Sierociuk, D., Malesza, W. and Macias, M. (2015). Numerical schemes for initialized constant and variable fractional-order derivatives: Matrix approach and its analog verification. *Journal of Vibration and Control*, 1–13. doi:10.1177/1077546314565438.

Valerio, D. and da Costa, J.S. (2011). Variable-order fractional derivatives and their numerical approximations. *Signal Processing*, 91(3), 470–483.

WHO Weight-for-age–World Health Organization. (2019). www.who.int/childgrowth/standards/weight_for_age/en/.

3

Fractional Calculus Approach in SIRS-SI Model for Malaria Disease with Mittag-Leffler Law

Jagdev Singh, Sunil Dutt Purohit and Devendra Kumar

CONTENTS

3.1 Introduction 83
3.2 The Basic Features of AB Fractional Derivative 85
3.3 SIRS-SI Malaria Disease Model Associated with Non-Local
    and Non-Singular Kernel 86
    3.3.1 Denotations 86
3.4 The EU Analysis of the Solution of the Fractional-Order SIRS-SI
    Malaria Disease Model 87
3.5 New Feature of q-HATM to Solve the SIRS-SI Malaria Disease
    Model of Fractional Nature 96
3.6 Numerical Results and Discussions 100
3.7 Conclusions 104
Acknowledgements 105
References 105

3.1 Introduction

Malaria is transmitted from one person to another person by a particular female mosquito, namely, the Anopheles mosquito, which is one of the most effective vectors for human disease. Several species have been discovered to be the vectors in various parts of the globe. In Africa, the chief vector is found to be *A. gambiae* complex and in the region of North America that is known as *A. freeborni*. In India, approximately 45 different species of the mosquito have been discovered. The main vectors that have been involved in the transmission of disease malaria are *A. fluviatilis*, *A. culicifacies*, *A. minimus*, *A. fluviatilis*, *A. stephensi*, *A. sundaicus*, and *A. philippinensis*. These mosquitoes are found to be in different areas. *A. minimus* and *A. fluviatilis* have been discovered in hilly regions of India, *A. sundaicus* and *A. stephensi* have appeared in the coastal areas, and *A. philippinensis* and *A. culicifacies* and have appeared in the plains regions. Children are the major victims of malaria disease. Children less than 5 years old have 57% of malaria diseases fatalities. Malaria is one of the major causes of child mortality. In 2017, every 12th child that died was due to malaria.

To get a better description about the spread and management of malaria diseases, a mathematical modelling of malaria is a very powerful tool. In recent years, several mathematicians and scientists have discussed mathematical models for malaria disease transmission. Abdullahi et al. [1] discussed mathematical modelling of malaria and the benefits of drugs. Onah et al. [2] studied the control measures for malaria diseases by applying a mathematical model. Huo and Qiu [3] represented the stability analysis of a mathematical model of malaria. Agusto et al. [4] studied usefulness of the optimal control for the malaria disease. Chiyaka and the co-authors [5] studied the mathematical representations qualitatively to establish the criteria to manage malaria disease. Mandal et al. [6] represented the mathematical models to understand about spreading and transmission of malaria diseases. Rafikov et al. [7] suggested an effective strategy for the malaria vector by utilising genetically moderated mosquitoes. Yang [8] discussed the malaria transmission related to the global warming, additionally with some local circumstances. Senthamarai et al. [9] applied a homotopy analysis technique for investigating the extending of malaria disease in the SIRS-SI model. Because of local features of classical order derivatives, all these discussed works and mathematical representations have their own limitations and problems. Hence, non-integer order derivative strategies are encouraged in mathematical modelling for biological and natural systems. In a very recent work, Kumar et al. [10] have suggested a advanced mathematical model to describe spreading and controlling of malaria disease by employing a novel fractional-order derivative having an exponential kernel in nature; for more details of other work see [11–23].

Recently, Atangana and Baleanu [24] suggested a fractional derivative, namely, the Atangana-Baleanu (AB) derivative with the Mittag-Leffler-type kernel. The AB derivative with non-integer order has been applied to represent several real-world problems, some of which Kumar et al. [25] studied about the new analysis of vibration equation with an arbitrary order derivative with strong memory effect. Kumar et al. [26] proposed new features of the Biswas-Milovic model having non-integer order derivative in optical communications. Singh et al. [27] discussed the regularised long-wave equation having an advanced fractional derivative. Kumar et al. [17] reported a new study of the Fornberg-Whitham equation associated with a fractional derivative in wave breaking. Singh [28] proposed a new rumour-spreading model pertaining to a fractional-order derivative in a social network and obtained very important and useful results for the public consequences and much more.

By observing the importance of fractional mathematical models in biology and effective features of the AB fractional derivative, we are encouraged to carry a new research work. We utilised the AB derivative in the SIRS-SI malaria disease model with a strong memory effect. With application of a fixed-point postulate, we obtained existence and uniqueness (EU) of the solution of the SIRS-SI malaria model representing spreading and controlling of malaria disease having a Mittag-Leffler-type memory. A novel numerical technique, the q-homotopy analysis transform method (q-HATM) [29,30], is applied to solve the fractional SIRS-SI malaria model with non-local and non-singular kernel. The newly developed q-HATM is a mixture of the q-homotopy analysis method (q-HAM) [31,32] and familiar Laplace transform approach [33–36]. This paper is constructed as: In Section 3.2, the fundamental characteristic of the AB derivative is discussed. Mathematical modelling of the SIRS-SI malaria model of fractional order is discussed in Section 3.3. Section 3.4 narrates the details of the EU of solution of the non-integer order SIRS-SI model describing the spreading and managing of malaria. In Section 3.5, the effectiveness of q-HATM is investigated to derive the solution of the SIRS-SI malaria disease model having fractional order. In Section 3.6, the consequences of several parameters on person-to-person and on different species of mosquitoes are analysed. And lastly, in Section 3.7, we discuss the conclusions.

3.2 The Basic Features of AB Fractional Derivative

Definition 3.1

Let us assume that $T \in H^1(e, f), f > e, \omega \in (0,1]$. Additionally, it is differentiable, then the AB non-integer order derivative in the sense of Caputo is defined in the following manner:

$$^{ABC}_{e}D^{\omega}_{\rho}(T(\rho)) = \frac{A(\omega)}{1-\omega} \int_{e}^{\rho} T'(\mu) E_{\omega}\left[-\frac{\omega}{1-\omega}(\rho-\mu)^{\omega}\right] d\mu. \quad (3.1)$$

In addition, $A(\omega)$ is fulfilling the features of $A(0) = A(1) = 1$.

Definition 3.2

Let us suppose $T \in H^1(e, f), f > e, \omega \in (0,1]$; moreover, it is non-differentiable, in this case the AB derivative in the Riemann-Liouville nature is defined as follows:

$$^{ABR}_{e}D^{\omega}_{\rho}(T(\rho)) = \frac{A(\omega)}{1-\omega} \frac{d}{d\rho} \int_{e}^{\rho} T(\mu) E_{\omega}\left[-\frac{\omega}{1-\omega}(\rho-\mu)^{\omega}\right] d\mu. \quad (3.2)$$

Definition 3.3

Let us assume that $0 < \omega < 1$, and T be a function of ρ in this case integral operator pertaining to AB derivative of fractional order ω is presented in the following way:

$${}^{AB}_{0}I^{\omega}_{\rho}\left(T(\rho)\right) = \frac{(1-\omega)}{A(\omega)}T(\rho) + \frac{\omega}{A(\omega)\Gamma(\omega)}\int_{0}^{\rho}T(\mu)(\rho-\mu)^{\omega-1}d\mu, \rho \geq 0. \qquad (3.3)$$

3.3 SIRS-SI Malaria Disease Model Associated with Non-Local and Non-Singular Kernel

Here, we transformed our compartmental malaria model (SIR-SI) into the systems of proportions of (sir-si) to describe the human populations and the mosquito populations by the systems of differential equations with ordinary derivatives.

3.3.1 Denotations

The variables s_h indicates the susceptible human; i_h denotes the infected human; r_h represents the recovered human; s_m denotes the susceptible mosquito; i_m stands for the infected mosquito; δ_h is the transfer rate of newly born humans to susceptible group; $\sigma\theta_1$ represents transfer rate of susceptible humans to infected group due to the blood transfusion; $\nu\theta_2$ indicates the infected mosquito's bite rate; $\lambda\theta_3$ denotes the transfer rate of susceptible mosquitoes to infected group due to bite the infected human; ψ stands for the vaccination rate; l_h, l_m are the symbols for the death rates of infected humans and mosquitoes, respectively; τ designates the congenital rate; $d\alpha$ shows the anti-malarial drug rate; η indicates the death rate due to malaria disease; δ_m symbolises the susceptible class rate with a constant; and β signifies the death rate of mosquitoes due to spraying.

Taking the nomenclature into account, the malaria model is represented as follows:

$$\frac{ds_h}{d\rho} = \delta_h + \gamma r_h - \left(\sigma\theta_1 i_h + \nu\theta_2 i_m\right)s_h - \left(\psi + l_h\right)s_h,$$

$$\frac{di_h}{d\rho} = \tau i_h + \left(\sigma\theta_1 i_h + \nu\theta_2 i_m\right)s_h - \left(l_h + \eta + d\alpha\right)i_h,$$

$$\frac{dr_h}{d\rho} = d\alpha i_h - \left(l_h + \gamma\right)r_h + \psi s_h, \qquad (3.4)$$

$$\frac{ds_m}{d\rho} = \delta_m - \left(\lambda\theta_3 i_h + l_m + \beta\right)s_m,$$

$$\frac{di_m}{d\rho} = \lambda\theta_3 i_h s_m - (l_m + \beta)i_m.$$

Since we know that the classical-order derivative does not describe the whole memory of the dynamic system, the discussed malaria disease model in equation (3.4) does not provide the complete details of the various consequences on person-to-person and on mosquitoes. So, by taking into consideration the complete memory of the malaria model, we develop the model in equation (3.4) by utilising the recently suggested AB fractional-order derivative in the following way:

$$^{ABC}D_0^\omega s_h = \delta_h + \gamma r_h - (\sigma\theta_1 i_h + v\theta_2 i_m)s_h - (\psi + l_h)s_h,$$

$$^{ABC}D_0^\omega i_h = \tau i_h + (\sigma\theta_1 i_h + v\theta_2 i_m)s_h - (l_h + \eta + d\alpha)i_h,$$

$$^{ABC}D_0^\omega r_h = d\alpha i_h - (l_h + \gamma)r_h + \psi s_h, \quad (3.5)$$

$$^{ABC}D_0^\omega s_m = \delta_m - (\lambda\theta_3 i_h + l_m + \beta)s_m,$$

$$^{ABC}D_0^\omega i_m = \lambda\theta_3 i_h s_m - (l_m + \beta)i_m,$$

having the initial conditions as

$$s_h(0) = d_1, i_h(0) = d_2, r_h(0) = d_3, s_m(0) = d_4, i_m(0) = d_5. \quad (3.6)$$

Let B symbolise the Banach space of a continuous $\Re \to \Re$ valued function specified on the interval I. Additionally, the related norm is as

$$\|(s_h, i_h, r_h, s_m, i_m)\| = \|s_h\| + \|i_h\| + \|r_h\| + \|s_m\| + \|i_m\|, \quad (3.7)$$

where $\|s_h\| = \sup\{|s_h(\rho)| : \rho \in I\}, \|i_h\| = \sup\{|i_h(\rho)| : \rho \in I\}, \|r_h\| = \sup\{|r_h(\rho)| : \rho \in I\},$ $\|s_m\| = \sup\{|s_m(\rho)| : \rho \in I\}$ and $\|i_m\| = \sup\{|i_m(\rho)| : \rho \in I\}$. Particularly $B = F(I) \times F(I)$, where $F(I)$ represents the Banach space of a continuous $\Re \to \Re$ valued functions on the interval I in addition related with the sup norm.

3.4 The EU Analysis of the Solution of the Fractional-Order SIRS-SI Malaria Disease Model

Here, we discuss EU of solution of the non-integer SIRS-SI malaria disease model with the Mittag-Leffler kernel. The analysis of the EU of the solution of any mathematical model arising in physical problems is very crucial. Hence, by employing the fixed-point theory, we discuss the EU of the solution of the arbitrary-order SIRS-SI malaria disease model.

First of all, we exert the integral operator of fractional order to the fractional malaria model in equation (3.5), and it yields

$$s_h(\rho) - s_h(0) = {}_0^{AB}I_\rho^\omega \{\delta_h + \gamma r_h - (\sigma\theta_1 i_h + v\theta_2 i_m)s_h - (\psi + l_h)s_h\},$$

$$i_h(\rho) - i_h(0) = {}_0^{AB}I_\rho^\omega \{\tau i_h + (\sigma\theta_1 i_h + v\theta_2 i_m)s_h - (l_h + \eta + d\alpha)i_h\}, \qquad (3.8)$$

$$r_h(\rho) - r_h(0) = {}_0^{AB}I_\rho^\omega \{d\alpha i_h - (l_h + \gamma)r_h + \psi s_h\},$$

$$s_m(\rho) - s_m(0) = {}_0^{AB}I_\rho^\omega \{\delta_m - (\lambda\theta_3 i_h + l_m + \beta)s_m\},$$

$$i_m(\rho) - i_m(0) = {}_0^{AB}I_\rho^\omega \{\lambda\theta_3 i_h s_m - (l_m + \beta)i_m\}.$$

By making use of equation (3.3), it converts in the following form:

$$s_h(\rho) - s_h(0) = \frac{(1-\omega)}{B(\omega)}\{\delta_h + \gamma r_h(\rho) - (\sigma\theta_1 i_h(\rho) + v\theta_2 i_m(\rho))s_h(\rho) - (\psi + l_h)s_h(\rho)\} +$$

$$\frac{\omega}{B(\omega)\Gamma(\omega)}\int_0^\rho \{\delta_h + \gamma r_h(\mu) - (\sigma\theta_1 i_h(\mu) + v\theta_2 i_m(\mu))s_h(\mu) - (\psi + l_h)s_h(\mu)\}(\rho-\mu)^{\omega-1}d\mu,$$

$$i_h(\rho) - i_h(0) = \frac{(1-\omega)}{B(\omega)}\{\tau i_h(\rho) + (\sigma\theta_1 i_h(\rho) + v\theta_2 i_m(\rho))s_h(\rho) - (l_h + \eta + d\alpha)i_h(\rho)\} +$$

$$\frac{\omega}{B(\omega)\Gamma(\omega)}\int_0^\rho \{\tau i_h(\mu) + (\sigma\theta_1 i_h(\mu) + v\theta_2 i_m(\mu))s_h(\mu) - (l_h + \eta + d\alpha)i_h(\mu)\}(\rho-\mu)^{\omega-1}d\mu,$$

$$r_h(\rho) - r_h(0) = \frac{(1-\omega)}{B(\omega)}\{d\alpha i_h(\rho) - (l_h + \gamma)r_h(\rho) + \psi s_h(\rho)\} +$$

$$\frac{\omega}{B(\omega)\Gamma(\omega)}\int_0^\rho \{d\alpha i_h(\mu) - (l_h + \gamma)r_h(\mu) + \psi s_h(\mu)\}(\rho-\mu)^{\omega-1}d\mu,$$

$$s_m(\rho) - s_m(0) = \frac{(1-\omega)}{B(\omega)}\{\delta_m - (\lambda\theta_3 i_h(\rho) + l_m + \beta)s_m(\rho)\} +$$

$$\qquad (3.9)$$

$$\frac{\omega}{B(\omega)\Gamma(\omega)}\int_0^\rho \{\delta_m - (\lambda\theta_3 i_h(\mu) + l_m + \beta)s_m(\mu)\}(\rho-\mu)^{\omega-1}d\mu,$$

$$i_m(\rho) - i_m(0) = \frac{(1-\omega)}{B(\omega)}\{\lambda\theta_3 i_h(\rho)s_m(\rho) - (l_m + \beta)i_m(\rho)\} +$$

$$\frac{\omega}{B(\omega)\Gamma(\omega)}\int_0^\rho \{\lambda\theta_3 i_h(\mu)s_m(\mu) - (l_m + \beta)i_m(\mu)\}(\rho-\mu)^{\omega-1}d\mu.$$

Fractional Calculus Approach in SIRS-SI Model

For clarity, we use the notations as

$$Y_1(\rho, s_h) = \delta_h + \gamma r_h(\rho) - (\sigma\theta_1 i_h(\rho) + v\theta_2 i_m(\rho)) s_h(\rho) - (\psi + l_h) s_h(\rho),$$

$$Y_2(\rho, i_h) = \tau i_h(\rho) + (\sigma\theta_1 i_h(\rho) + v\theta_2 i_m(\rho)) s_h(\rho) - (l_h + \eta + d\alpha) i_h(\rho),$$

$$Y_3(\rho, r_h) = d\alpha i_h(\rho) - (l_h + \gamma) r_h(\rho) + \psi s_h(\rho), \tag{3.10}$$

$$Y_4(\rho, s_m) = \delta_m - (\lambda\theta_3 i_h(\rho) + l_m + \beta) s_m(\rho),$$

$$Y_5(\rho, i_m) = \lambda\theta_3 i_h(\rho) s_m(\rho) - (l_m + \beta) i_m(\rho).$$

Theorem 3.1

The kernels $Y_1(\rho, s_h), Y_2(\rho, i_h), Y_3(\rho, r_h), Y_4(\rho, s_m)$ and $Y_5(\rho, i_m)$ fulfil the Lipchitz condition and contraction if the following inequality is satisfied:

$$0 \leq \omega\gamma_1 a_2 + \xi\gamma_2 a_5 + (\delta + \eta_h) < 1.$$

Proof

We start with $Y_1(\rho, s_h)$. Let us assume that s_h and s_{h1} are two functions, then we get the following result:

$$\|Y_1(\rho, s_h) - Y_1(\rho, s_{h1})\|$$
$$= \|-(\sigma\theta_1 i_h(\rho) + v\theta_2 i_m(\rho))\{s_h(\rho) - s_{h1}(\rho)\} - (\psi + l_h)\{s_h(\rho) - s_{h1}(\rho)\}\|. \tag{3.11}$$

By exerting rules of norm on equation (3.11), it provides

$$\|Y_1(\rho, s_h) - Y_1(\rho, s_{h1})\| \leq \|\{s_h(\rho) - s_{h1}(\rho)\}(\sigma\theta_1 i_h(\rho) + v\theta_2 i_m(\rho))\|$$
$$+ \|\{s_h(\rho) - s_{h1}(\rho)\}(\psi + l_h)\|$$
$$\leq \{\sigma\theta_1 \|i_h(\rho)\| + v\theta_2 \|i_m(\rho)\| + (\psi + l_h)\} \|s_h(\rho) - s_{h1}(\rho)\|$$
$$\leq \{\sigma\theta_1 b_2 + v\theta_2 b_5 + (\psi + l_h)\} \|s_h(\rho) - s_{h1}(\rho)\|$$
$$\leq c_1 \|s_h(\rho) - s_{h1}(\rho)\|. \tag{3.12}$$

Substituting $c_1 = \sigma\theta_1 b_2 + v\theta_2 b_5 + (\psi + l_h)$ where $\|s_h(\rho)\| \leq b_1, \|i_h(\rho)\| \leq b_2, \|r_h(\rho)\| \leq b_3,$ $\|s_m(\rho)\| \leq b_4$ and $\|i_m(\rho)\| \leq b_5$ are the bounded functions in nature, then, we get

$$\|Y_1(\rho, s_h) - Y_1(\rho, s_{h1})\| \leq c_1 \|s_h(\rho) - s_{h1}(\rho)\|. \tag{3.13}$$

Thus, the Lipschiz condition is satisfied for Y_1. Besides, if $0 \leq \sigma\theta_1 b_2 + v\theta_2 b_5 + (\psi + l_h) < 1$, then it is also a contraction.

In the same way, we can verify that the kernels $Y_2(\rho, i_h), Y_3(\rho, r_h), Y_4(\rho, s_m)$ and $Y_5(\rho, i_m)$ fulfil the Lipschiz conditions and are given as follows:

$$\|Y_2(\rho, i_h) - Y_2(\rho, i_{h1})\| \leq c_2 \|i_h(\rho) - i_{h1}(\rho)\|,$$

$$\|Y_3(\rho, r_h) - Y_3(\rho, r_{h1})\| \leq c_3 \|r_h(\rho) - r_{h1}(\rho)\|, \qquad (3.14)$$

$$\|Y_4(\rho, s_m) - Y_4(\rho, s_{m1})\| \leq c_4 \|s_m(\rho) - s_{m1}(\rho)\|,$$

$$\|Y_5(\rho, i_m) - Y_5(\rho, i_{m1})\| \leq c_5 \|i_m(\rho) - i_{m1}(\rho)\|.$$

By making use of the symbols of the initially discussed kernels, then equation (3.9) is converted to subsequent system as

$$s_h(\rho) = s_h(0) + \frac{(1-\omega)}{B(\omega)} Y_1(\rho, s_h) + \frac{\omega}{B(\omega)\Gamma(\omega)} \int_0^\rho Y_1(\mu, s_h)(\rho - \mu)^{\omega-1} d\mu,$$

$$i_h(\rho) = i_h(0) + \frac{(1-\omega)}{B(\omega)} Y_2(\rho, i_h) + \frac{\omega}{B(\omega)\Gamma(\omega)} \int_0^\rho Y_2(\mu, i_h)(\rho - \mu)^{\omega-1} d\mu,$$

$$r_h(\rho) = r_h(0) + \frac{(1-\omega)}{B(\omega)} Y_3(\rho, i_h) + \frac{\omega}{B(\omega)\Gamma(\omega)} \int_0^\rho Y_3(\mu, r_h)(\rho - \mu)^{\omega-1} d\mu,$$

$$s_m(\rho) = s_m(0) + \frac{(1-\omega)}{B(\omega)} Y_4(\rho, s_m) + \frac{\omega}{B(\omega)\Gamma(\omega)} \int_0^\rho Y_4(\mu, s_h)(\rho - \mu)^{\omega-1} d\mu, \quad (3.15)$$

$$i_m(\rho) = i_m(0) + \frac{(1-\omega)}{B(\omega)} Y_5(\rho, i_m) + \frac{\omega}{B(\omega)\Gamma(\omega)} \int_0^\rho Y_5(\mu, i_m)(\rho - \mu)^{\omega-1} d\mu.$$

Then, we develop the subsequent recursive formula

$$s_{hn}(\rho) = \frac{(1-\omega)}{A(\omega)} Y_1(\rho, s_{h(n-1)}) + \frac{\omega}{A(\omega)\Gamma(\omega)} \int_0^\rho Y_1(\mu, s_{h(n-1)})(\rho - \mu)^{\omega-1} d\mu,$$

$$i_{hn}(\rho) = \frac{(1-\omega)}{A(\omega)} Y_2(\rho, i_{h(n-1)}) + \frac{\omega}{A(\omega)\Gamma(\omega)} \int_0^\rho Y_2(\mu, i_{h(n-1)})(\rho - \mu)^{\omega-1} d\mu,$$

$$r_{hn}(\rho) = \frac{(1-\omega)}{A(\omega)} Y_3(\rho, r_{h(n-1)}) + \frac{\omega}{A(\omega)\Gamma(\omega)} \int_0^\rho Y_3(\mu, r_{h(n-1)})(\rho-\mu)^{\omega-1} d\mu,$$

$$S_{mn}(\rho) = \frac{(1-\omega)}{A(\omega)} Y_4(\rho, s_{m(n-1)}) + \frac{\omega}{A(\omega)\Gamma(\omega)} \int_0^\rho Y_4(\mu, s_{m(n-1)})(\rho-\mu)^{\omega-1} d\mu, \quad (3.16)$$

$$i_{mn}(\rho) = \frac{(1-\omega)}{A(\omega)} Y_5(\rho, i_{m(n-1)}) + \frac{\omega}{A(\omega)\Gamma(\omega)} \int_0^\rho Y_5(\mu, i_{m(n-1)})(\rho-\mu)^{\omega-1} d\mu.$$

having the initial conditions

$$s_{h0} = s_h(0),\ i_{h0} = i_h(0),\ r_{h0} = r_h(0),\ s_{m0} = s_m(0),\ i_{m0} = i_m(0). \quad (3.17)$$

The difference rule represented by the symbols is written as follows:

$$E_{1n}(\rho) = s_{hn}(\rho) - s_{h(n-1)}(\rho) = \frac{(1-\omega)}{A(\omega)} \left(Y_1(\rho, s_{h(n-1)}) - Y_1(\rho, s_{h(n-2)})\right)$$

$$+ \frac{\omega}{A(\omega)\Gamma(\omega)} \int_0^\rho \left(Y_1(\mu, s_{h(n-1)}) - Y_1(\mu, s_{h(n-2)})\right)(\rho-\mu)^{\omega-1} d\mu,$$

$$E_{2n}(\rho) = i_{hn}(\rho) - i_{h(n-1)}(\rho) = \frac{(1-\omega)}{A(\omega)} \left(Y_2(\rho, i_{h(n-1)}) - Y_2(\rho, i_{h(n-2)})\right)$$

$$+ \frac{\omega}{A(\omega)\Gamma(\omega)} \int_0^\rho \left(Y_2(\mu, i_{h(n-1)}) - Y_2(\mu, i_{h(n-2)})\right)(\rho-\mu)^{\omega-1} d\mu,$$

$$E_{3n}(\rho) = r_{hn}(\rho) - r_{h(n-1)}(\rho) = \frac{(1-\omega)}{A(\omega)} \left(Y_3(\rho, r_{h(n-1)}) - Y_3(\rho, r_{h(n-2)})\right)$$

$$+ \frac{\omega}{A(\omega)\Gamma(\omega)} \int_0^\rho \left(Y_3(\mu, r_{h(n-1)}) - Y_3(\mu, r_{h(n-2)})\right)(\rho-\mu)^{\omega-1} d\mu,$$

$$E_{4n}(\rho) = s_{mn}(\rho) - s_{m(n-1)}(\rho) = \frac{(1-\omega)}{A(\omega)} \left(Y_4(\rho, s_{m(n-1)}) - Y_4(\rho, s_{m(n-2)})\right)$$

$$+ \frac{\omega}{A(\omega)\Gamma(\omega)} \int_0^\rho \left(Y_4(\mu, s_{m(n-1)}) - Y_4(\mu, s_{m(n-2)})\right)(\rho-\mu)^{\omega-1} d\mu,$$

$$(3.18)$$

$$E_{5n}(\rho) = i_{mn}(\rho) - i_{m(n-1)}(\rho) = \frac{(1-\omega)}{A(\omega)}\left(Y_5(\rho, i_{m(n-1)}) - Y_5(\rho, i_{m(n-2)})\right)$$

$$+ \frac{\omega}{A(\omega)\Gamma(\omega)} \int_0^\rho \left(Y_5(\mu, i_{m(n-1)}) - Y_5(\mu, i_{m(n-2)})\right)(\rho - \mu)^{\omega-1} d\mu.$$

It is crucial to notice that

$$s_{hn}(\rho) = \sum_{j=0}^n E_{1j}(\rho),\ i_{hn}(\rho) = \sum_{j=0}^n E_{2j}(\rho),\ r_{hn}(\rho) = \sum_{j=0}^n E_{3j}(\rho),$$

$$s_{mn}(\rho) = \sum_{j=0}^n E_{4j}(\rho),\ i_{mn}(\rho) = \sum_{j=0}^n E_{5j}(\rho). \tag{3.19}$$

Now, we can easily obtain the subsequent result

$$\|E_{1n}(\rho)\| = \|s_{hn}(\rho) - s_{h(n-1)}(\rho)\| = \left\| \begin{array}{c} \frac{(1-\omega)}{A(\omega)}\left(Y_1(\rho, s_{h(n-1)}) - Y_1(\rho, s_{h(n-2)})\right) + \frac{\omega}{A(\omega)\Gamma(\omega)} \\ \int_0^\rho \left(Y_1(\mu, s_{h(n-1)}) - Y_1(\mu, s_{h(n-2)})\right)(\rho-\mu)^{\omega-1} d\mu \end{array} \right\|.$$

$$\tag{3.20}$$

By utilising triangular inequality on equation (3.20), we obtain the following results:

$$\|s_{hn}(\rho) - s_{h(n-1)}(\rho)\| \le \frac{(1-\omega)}{A(\omega)} \|Y_1(\rho, s_{h(n-1)}) - Y_1(\rho, s_{h(n-2)})\|$$

$$+ \frac{\omega}{A(\omega)\Gamma(\omega)} \left\| \int_0^\rho \left(Y_1(\mu, s_{h(n-1)}) - Y_1(\mu, s_{h(n-2)})\right)(\rho-\mu)^{\omega-1} d\mu \right\|.$$

$$\tag{3.21}$$

We have already verified that $Y_1(\rho, s_h)$ possesses the Lipchitz condition, thus, we have

$$\|s_{hn}(\rho) - s_{h(n-1)}(\rho)\| \le \frac{(1-\omega)}{A(\omega)} c_1 \|s_{h(n-1)} - s_{h(n-2)}\|$$

$$+ \frac{\omega}{A(\omega)\Gamma(\omega)} c_1 \int_0^\rho \|s_{h(n-1)} - s_{h(n-2)}\|(\rho-\mu)^{\omega-1} d\mu,$$

$$\tag{3.22}$$

Fractional Calculus Approach in SIRS-SI Model

Hence, we come at the following result:

$$\|E_{1n}(\rho)\| \leq \frac{(1-\omega)}{A(\omega)} c_1 \|E_{1(n-1)}(\rho)\| + \frac{\omega}{A(\omega)\Gamma(\omega)} c_1 \int_0^\rho \|E_{1(n-1)}(\rho)\|(\rho-\mu)^{\omega-1} d\mu. \quad (3.23)$$

In the similar process, we obtain as follows:

$$\|E_{2n}(\rho)\| \leq \frac{(1-\omega)}{A(\omega)} c_2 \|E_{2(n-1)}(\rho)\| + \frac{\omega}{A(\omega)\Gamma(\omega)} c_2 \int_0^\rho \|E_{2(n-1)}(\rho)\|(\rho-\mu)^{\omega-1} d\mu,$$

$$\|E_{3n}(\rho)\| \leq \frac{(1-\omega)}{A(\omega)} c_3 \|E_{3(n-1)}(\rho)\| + \frac{\omega}{A(\omega)\Gamma(\omega)} c_3 \int_0^\rho \|E_{3(n-1)}(\rho)\|(\rho-\mu)^{\omega-1} d\mu, \quad (3.24)$$

$$\|E_{4n}(\rho)\| \leq \frac{(1-\omega)}{A(\omega)} c_4 \|E_{4(n-1)}(\rho)\| + \frac{\omega}{A(\omega)\Gamma(\omega)} c_4 \int_0^\rho \|E_{4(n-1)}(\rho)\|(\rho-\mu)^{\omega-1} d\mu,$$

$$\|E_{5n}(\rho)\| \leq \frac{(1-\omega)}{A(\omega)} c_5 \|E_{3(n-1)}(\rho)\| + \frac{\omega}{A(\omega)\Gamma(\omega)} c_5 \int_0^\rho \|E_{5(n-1)}(\rho)\|(\rho-\mu)^{\omega-1} d\mu.$$

In taking into consideration the outcomes of equations (3.23) and (3.24), we verify the existence of the solution of the malaria disease model (equation 3.5).

Theorem 3.2

The discussed mathematical model given in equation (3.5) yields a solution if we can find a value ρ_0 such that

$$\frac{(1-\omega)}{A(\omega)} c_1 + \frac{\omega}{A(\omega)\Gamma(\omega+1)} c_1 \rho_0^\omega < 1.$$

Proof

Taking the results of equations (3.23) and (3.24), we have

$$\|E_{1n}(\rho)\| \leq \|s_{hn}(0)\| \left[\left(\frac{(1-\omega)}{A(\omega)} c_1\right) + \left(\frac{\omega}{A(\omega)\Gamma(\omega+1)} c_1 \rho^\omega\right)\right]^n,$$

$$\|E_{2n}(\rho)\| \leq \|i_{hn}(0)\| \left[\left(\frac{(1-\omega)}{A(\omega)} c_2\right) + \left(\frac{\omega}{A(\omega)\Gamma(\omega+1)} c_2 \rho^\omega\right)\right]^n, \quad (3.25)$$

$$\|E_{3n}(\rho)\| \leq \|r_{hn}(0)\| \left[\left(\frac{(1-\omega)}{A(\omega)} c_3 \right) + \left(\frac{\omega}{A(\omega)\Gamma(\omega+1)} c_3 \rho^\omega \right) \right]^n,$$

$$\|E_{4n}(\rho)\| \leq \|s_{mn}(0)\| \left[\left(\frac{(1-\omega)}{A(\omega)} c_4 \right) + \left(\frac{\omega}{A(\omega)\Gamma(\omega+1)} c_4 \rho^\omega \right) \right]^n,$$

$$\|E_{5n}(\rho)\| \leq \|i_{hn}(0)\| \left[\left(\frac{(1-\omega)}{A(\omega)} c_5 \right) + \left(\frac{\omega}{A(\omega)\Gamma(\omega+1)} c_5 \rho^\omega \right) \right]^n.$$ Therefore, the previously stated solutions exist and are continuous. Now, in order to show that equation (3.16) is a solution of equation (3.5), we put

$$s_h(\rho) - s_h(0) = s_{hn}(\rho) - G_{1n}(\rho),$$

$$i_h(\rho) - i_h(0) = i_{hn}(\rho) - G_{2n}(\rho),$$

$$r_h(\rho) - r_h(0) = r_{hn}(\rho) - G_{3n}(\rho), \quad (3.26)$$

$$s_m(\rho) - s_m(0) = s_{mn}(\rho) - G_{4n}(\rho),$$

$$i_m(\rho) - i_m(0) = i_{mn}(\rho) - G_{5n}(\rho).$$

Hence, we have the following result:

$$\|G_{1n}(\rho)\| = \left\| \frac{(1-\omega)}{A(\omega)} \left(Y_1(\rho, s_h) - Y_1(\rho, s_{h(n-1)}) \right) \right.$$
$$\left. + \frac{\omega}{A(\omega)\Gamma(\omega)} \int_0^\rho \left(Y_1(\mu, s_h) - Y_1(\mu, s_{h(n-1)}) \right)(\rho-\mu)^{\omega-1} d\mu \right\|$$

$$\leq \frac{(1-\omega)}{A(\omega)} \|Y_1(\rho, s_h) - Y_1(\rho, s_{h(n-1)})\|$$

$$+ \frac{\omega}{A(\omega)\Gamma(\omega)} \int_0^\rho \|Y_1(\mu, s_h) - Y_1(\mu, s_{h(n-1)})\| (\rho-\mu)^{\omega-1} d\mu,$$

$$\leq \frac{(1-\omega)}{A(\omega)} c_1 \|s(\rho) - s_{h(n-1)}(\rho)\| + \frac{\omega}{A(\omega)\Gamma(\omega+1)} c_1 \|s(\rho) - s_{h(n-1)}(\rho)\| \rho^\omega. \quad (3.27)$$

By making use of this processor recursively, it yields

$$\|G_{1n}(\rho)\| \leq \left(\frac{(1-\omega)}{A(\omega)} + \frac{\omega}{A(\omega)\Gamma(\omega+1)} \rho^\omega \right)^{n+1} c_1^{n+1} b_1. \quad (3.28)$$

Thus, at the value $\rho = \rho_0$, we get

$$\|G_{1n}(\rho)\| \leq \left(\frac{(1-\omega)}{A(\omega)} + \frac{\omega}{A(\omega)\Gamma(\omega+1)}\rho_0^\omega\right)^{n+1} c_1^{n+1} b_1. \qquad (3.29)$$

Subsequently, when we take limit $n \to \infty$ on equation (3.29), it yields

$$\|G_{1n}(t)\| \to 0.$$

In the similar processor, we have
$\|G_{2n}(t)\| \to 0, \|G_{3n}(t)\| \to 0, \|G_{4n}(t)\| \to 0$ and $\|G_{5n}(t)\| \to 0$.

Hence, we verified the existence of solution of the model in equation (3.5).

Afterwards, we prove the uniqueness of solution of the malaria model in equation (3.5).

We assume that there exists one more solution of the model in equation (3.5), that is, $s_h^*(\rho), i_h^*(\rho), r_h^*(\rho), s_m^*(\rho)$ and $i_m^*(\rho)$ then, we have

$$s_h(\rho) - s_h^*(\rho) = \frac{(1-\omega)}{A(\omega)}\left(Y_1(\rho, s_h) - Y_1(\rho, s_h^*)\right) + \\ \frac{\omega}{A(\omega)\Gamma(\omega)}\int_0^\rho \left(Y_1(\mu, s_h) - Y_1(\mu, s_h^*)\right)(\rho - \mu)^{\omega-1} d\mu. \qquad (3.30)$$

Now, by utilising the norm on both sides of equation (3.30), it provides

$$\|s_h(\rho) - s_h^*(\rho)\| \leq \frac{(1-\omega)}{A(\omega)}\|Y_1(\rho, s_h) - Y_1(\rho, s_h^*)\| + \\ \frac{\omega}{A(\omega)\Gamma(\omega)}\int_0^\rho \|(Y_1(\mu, s_h) - Y_1(\mu, s_h^*))\|(\rho - \mu)^{\omega-1} d\mu. \qquad (3.31)$$

By making use of Lipschitz condition of $Y_1(\rho, s_h)$, we get

$$\|s_h(\rho) - s_h^*(\rho)\| \leq \frac{(1-\omega)}{A(\omega)} c_1 \|s_h(\rho) - s_h^*(\rho)\| + \frac{\rho^\omega}{A(\omega)\Gamma(\omega)} c_1 \|s_h(\rho) - s_h^*(\rho)\|. \qquad (3.32)$$

It yields the results

$$\|s_h(\rho) - s_h^*(\rho)\|\left(1 - \frac{(1-\omega)}{A(\omega)} c_1 - \frac{\rho^\omega}{A(\omega)\Gamma(\omega)} c_1\right) \leq 0. \qquad (3.33)$$

Theorem 3.3

If the result given by equation (3.34) is fulfilled, then the malaria model with non-integer order (equation 3.5) has a unique solution

$$\left(1 - \frac{(1-\omega)}{A(\omega)} c_1 - \frac{\rho^\omega}{A(\omega)\Gamma(\omega)} c_1\right) > 0. \qquad (3.34)$$

Proof

By taking the results of equation (3.33), we have

$$\left\| s_h(\rho) - s_h^*(\rho) \right\| \left(1 - \frac{(1-\omega)}{A(\omega)} c_1 - \frac{\rho^\omega}{A(\omega)\Gamma(\omega)} c_1 \right) \leq 0. \qquad (3.35)$$

After that, when we use the result of equation (3.34) and then the important characteristic of the norm in the result of equation (3.35), it provides

$$\left\| s_h(\rho) - s_h^*(\rho) \right\| = 0.$$

Hence, we have

$$s_h(\rho) = s_h^*(\rho). \qquad (3.36)$$

In the same way, we verify that

$$i_h = i_h^*,\ r_h = r_h^*,\ s_m = s_m^*,\ i_m = i_m^*. \qquad (3.37)$$

In this way, we have proved that the SIRS-SI malaria disease model having a fractional-order derivative (equation 3.5) has a unique solution.

3.5 New Feature of q-HATM to Solve the SIRS-SI Malaria Disease Model of Fractional Nature

Here to get the better biological description and the numerical outcomes of the model in equation (3.5), we will employ the newly method popularly known as q-HATM [29,30]. First, we take the Laplace transform on both sides of the model (equation 3.5), and after simplifying, we have

$$L[s_h] - \frac{d_1}{q} - \frac{1}{A(\omega)} \frac{\left[q^\omega + \omega(1 - q^\omega) \right] \delta_h}{q^{\omega+1}}$$

$$- \frac{1}{A(\omega)} \frac{q^\omega + \omega(1 - q^\omega)}{q^\omega} L\left[\gamma r_h - (\sigma \theta_1 i_h + v \theta_2 i_m) s_h - (\psi + l_h) s_h \right] = 0,$$

$$L[i_h] - \frac{d_2}{q} - \frac{1}{A(\omega)} \frac{q^\omega + \omega(1 - q^\omega)}{q^\omega} L\left[\tau i_h + (\sigma \theta_1 i_h + v \theta_2 i_m) s_h - (l_h + \eta + d\alpha) i_h \right] = 0,$$

$$L[r_h] - \frac{d_3}{q} - \frac{1}{A(\omega)} \frac{q^\omega + \omega(1 - q^\omega)}{q^\omega} L\left[d\alpha i_h - (l_h + \gamma) r_h + \psi s_h \right] = 0, \qquad (3.38)$$

$$L[s_m] - \frac{d_4}{q} - \frac{1}{A(\omega)} \frac{\left[q^\omega + \omega(1-q^\omega)\right]\delta_m}{q^{\omega+1}}$$

$$- \frac{1}{A(\omega)} \frac{q^\omega + \omega(1-q^\omega)}{q^\omega} L\left[-(\lambda\theta_3 i_h + l_m + \beta)s_m\right] = 0,$$

$$L[i_m] - \frac{d_5}{q} - \frac{1}{A(\omega)} \frac{q^\omega + \omega(1-q^\omega)}{q^\omega} L\left[\lambda\theta_3 i_h s_m - (l_m + \beta)i_m\right] = 0.$$

Next, the non-linear operators are given in the following manner:

$$H_1\left[\Lambda_1(\rho;y)\right] = L\left[\Lambda_1(\rho;y)\right] - \frac{d_1}{q} - \frac{1}{A(\omega)} \frac{\left[q^\omega + \omega(1-q^\omega)\right]\delta_h}{q^{\omega+1}}$$

$$- \frac{1}{A(\omega)} \frac{q^\omega + \omega(1-q^\omega)}{q^\omega} L[\gamma\Lambda_3(\rho;y)]$$

$$-(\sigma\theta_1\Lambda_2(\rho;y) + \upsilon\theta_2\Lambda_5(\rho;y))\Lambda_1(\rho;y) - (\psi + l_h)\Lambda_1(\rho;y)\Big] = 0,$$

$$H_2\left[\Lambda_2(\rho;y)\right] = L\left[\Lambda_2(\rho;y)\right] - \frac{d_2}{q} - \frac{1}{A(\omega)} \frac{q^\omega + \omega(1-q^\omega)}{q^\omega} L[\tau\Lambda_2(\rho;y) + (\sigma\theta_1\Lambda_2(\rho;y)$$

$$+ \upsilon\theta_2\Lambda_5(\rho;y))\Lambda_1(\rho;y) - (l_h + \eta + d\alpha)\Lambda_2(\rho;y)] = 0,$$

$$H_3\left[\Lambda_3(\rho;y)\right] = L\left[\Lambda_3(\rho;y)\right] - \frac{d_3}{q} - \frac{1}{A(\omega)} \frac{q^\omega + \omega(1-q^\omega)}{q^\omega} L[d\alpha\Lambda_2(\rho;y) - \quad (3.39)$$

$$(l_h + \gamma)\Lambda_3(\rho;y) + \psi\Lambda_1(\rho;y)] = 0,$$

$$H_4\left[\Lambda_4(\rho;y)\right] = L\left[\Lambda_4(\rho;y)\right] - \frac{d_4}{q} - \frac{1}{A(\omega)} \frac{\left[q^\omega + \omega(1-q^\omega)\right]\delta_m}{q^{\omega+1}} - \frac{1}{A(\omega)} \frac{q^\omega + \omega(1-q^\omega)}{q^\omega}$$

$$L[-(\lambda\theta_3\Lambda_2(\rho;y) + l_m + \beta)\Lambda_4(\rho;y)] = 0,$$

$$H_5\left[\Lambda_5(\rho;y)\right] = L\left[\Lambda_5(\rho;y)\right] - \frac{d_5}{q} - \frac{1}{A(\omega)} \frac{q^\omega + \omega(1-q^\omega)}{q^\omega} L[\lambda\theta_3\Lambda_2(\rho;y)\Lambda_4(\rho;y)$$

$$- (l_m + \beta)\Lambda_5(\rho;y)] = 0.$$

Hence, we have the following results:

$$\Omega_{1,u}\left(\vec{s}_{h(u-1)}\right) = L[s_{h(u-1)}] - \left(\frac{d_1}{q} + \frac{1}{A(\omega)} \frac{q^\omega + \omega(1-q^\omega)\delta_h}{q^{\omega+1}}\right)\left(1 - \frac{\chi_u}{n}\right)$$

$$-\frac{1}{A(\omega)} \frac{q^\omega + \omega(1-q^\omega)}{q^\omega} L\left[\gamma r_{h(u-1)} - \sigma\theta_1\left(\sum_{v=0}^{u-1} i_{hv} s_{h(u-1-v)}\right)\right.$$

$$\left. -\upsilon\theta_2\left(\sum_{v=0}^{u-1} i_{mv} s_{h(u-1-v)}\right) - (\psi + l_h) s_{h(u-1)}\right],$$

$$\Omega_{2,u}\left(\vec{i}_{h(u-1)}\right) = L[i_{h(u-1)}] - \frac{d_2}{q}\left(1 - \frac{\chi_u}{n}\right)$$

$$-\frac{1}{A(\omega)} \frac{q^\omega + \omega(1-q^\omega)}{q^\omega} L\left[\tau i_{h(u-1)} + \sigma\theta_1\left(\sum_{v=0}^{u-1} i_{hv} s_{h(u-1-v)}\right)\right.$$

$$\left. +\upsilon\theta_2\left(\sum_{v=0}^{u-1} i_{mv} s_{h(u-1-v)}\right) - (l_h + \eta + d\alpha) i_{h(u-1)}\right],$$

$$\Omega_{3,u}\left(\vec{r}_{h(u-1)}\right) = L[r_{h(u-1)}] - \frac{d_3}{q}\left(1 - \frac{\chi_u}{n}\right) \quad (3.40)$$

$$-\frac{1}{A(\omega)} \frac{q^\omega + \omega(1-q^\omega)}{q^\omega} L\left[d\alpha i_{h(u-1)} - (l_h + \gamma) r_{h(u-1)} + \psi s_{h(u-1)}\right],$$

$$\Omega_{4,u}\left(\vec{s}_{m(u-1)}\right) = L[s_{m(u-1)}] - \left(\frac{d_4}{q} + \frac{1}{A(\omega)} \frac{[q^\omega + \omega(1-q^\omega)]\delta_m}{q^{\omega+1}}\right)\left(1 - \frac{\chi_u}{n}\right)$$

$$-\frac{1}{A(\omega)} \frac{q^\omega + \omega(1-q^\omega)}{q^\omega} L\left[-\lambda\theta_3\left(\sum_{u=0}^{v-1} i_{hv} s_{m(u-1-v)}\right) - (l_m + \beta) s_{m(u-1)}\right],$$

$$\Omega_{5,u}\left(\vec{i}_{m(u-1)}\right) = L[i_{m(u-1)}] - \frac{d_5}{q}\left(1 - \frac{\chi_u}{n}\right)$$

$$-\frac{1}{A(\omega)} \frac{q^\omega + \omega(1-q^\omega)}{q^\omega} L\left[\lambda\theta_3\left(\sum_{u=0}^{v-1} i_{hu} s_{m(u-1-v)}\right) - (l_m + \beta) i_{m(u-1)}\right].$$

Now, the uth-order deformation equations are represented in the following manner:

$$L[s_{hu}(\rho) - \chi_u s_{h(u-1)}(\rho)] = \hbar \Omega_{1,u}\left(\vec{s}_{h(u-1)}\right),$$

$$L[i_{hu}(\rho) - \chi_u i_{h(u-1)}(\rho)] = \hbar \Omega_{2,u}\left(\vec{i}_{h(u-1)}\right),$$

$$L[r_{hu}(\rho) - \chi_u r_{h(u-1)}(\rho)] = \hbar \Omega_{3,u}\left(\vec{r}_{h(u-1)}\right),$$

$$L[s_{mu}(\rho) - \chi_u s_{m(u-1)}(\rho)] = \hbar \Omega_{4,u}\left(\vec{s}_{m(u-1)}\right), \quad (3.41)$$

$$L[i_{mu}(\rho) - \chi_u i_{m(u-1)}(\rho)] = \hbar \Omega_{5,u}\left(\vec{i}_{m(u-1)}\right).$$

By using the inverse Laplace transform on both the sides of equation (3.41), it gives

$$s_{hu}(\rho) = \chi_u s_{h(u-1)}(\rho) + \hbar L^{-1}\left[\Omega_{1,u}\left(\vec{s}_{h(u-1)}\right)\right],$$

$$i_{hu}(\rho) = \chi_u i_{h(u-1)}(\rho) + \hbar L^{-1}\left[\Omega_{2,u}\left(\vec{i}_{h(u-1)}\right)\right],$$

$$r_{hu}(\rho) = \chi_u r_{h(u-1)}(\rho) + \hbar L^{-1}\left[\Omega_{3,u}\left(\vec{r}_{h(u-1)}\right)\right], \quad (3.42)$$

$$s_{mu}(\rho) = \chi_u r_{m(u-1)}(\rho) + \hbar L^{-1}\left[\Omega_{4,u}\left(\vec{s}_{m(u-1)}\right)\right],$$

$$i_{mu}(\rho) = \chi_u i_{m(u-1)}(\rho) + \hbar L^{-1}\left[\Omega_{5,u}\left(\vec{i}_{m(u-1)}\right)\right].$$

By taking into consideration the initial guess, $s_{h0}(\rho) = d_1 + \left\{1 + \omega \dfrac{1}{A(\omega)}\left(\dfrac{\rho^\omega}{\Gamma(\omega+1)} - 1\right)\right\}\delta_h,\ \ i_{h0}(\rho) = d_2,\ \ r_{h0}(\rho) = d_3,$

$s_{m0}(\rho) = d_4 + \left\{1 + \omega \dfrac{1}{A(\omega)}\left(\dfrac{\rho^\omega}{\Gamma(\omega+1)} - 1\right)\right\}\delta_m,\ \ i_{m0}(\rho) = d_5$ and now for solving equation (3.42) for the values of $u = 1,2,3,\ldots$, we get $s_{hu}(\rho), i_{hu}(\rho), r_{hu}(\rho), s_{mu}(\rho)$ and $i_{mu}(\rho), \forall u \geq 1$.

Hence, the solution of the model (equation 3.5) is represented in the following way:

$$s_h(\rho) = s_{h0}(\rho) + s_{h1}(\rho)\left(\frac{1}{n}\right) + s_{h2}(\rho)\left(\frac{1}{n}\right)^2 + \ldots$$

$$i_h(\rho) = i_{h0}(\rho) + i_{h1}(\rho)\left(\frac{1}{n}\right) + i_{h2}(\rho)\left(\frac{1}{n}\right)^2 + \ldots$$

$$r_h(\rho) = r_{h0}(\rho) + r_{h1}(\rho)\left(\frac{1}{n}\right) + r_{h2}(\rho)\left(\frac{1}{n}\right)^2 + \ldots \quad (3.43)$$

$$s_m(\rho) = s_{m0}(\rho) + s_{m1}(\rho)\left(\frac{1}{n}\right) + s_{m2}(\rho)\left(\frac{1}{n}\right)^2 + \ldots$$

$$i_m(\rho) = i_{m0}(\rho) + i_{m1}(\rho)\left(\frac{1}{n}\right) + i_{m2}(\rho)\left(\frac{1}{n}\right)^2 + \ldots.$$

3.6 Numerical Results and Discussions

In this section, we compute the numerical results for the fractional-order mathematical model (equation 3.5). With the aid of q-HATM, the numerical results for the model (equation 3.5) are investigated. The numerical results are demonstrated at $l_h = 0.0004$, $l_m = 0.04$, $v = 0.05$, $\lambda = 1/730$, $\beta = 1/730$, $\delta_h = 0.027$, $\delta_m = 0.13$, $\sigma = 0.038$, $\eta = 0.13$, $\theta_1 = 0.02$, $\theta_2 = 0.010$, $\theta_3 = 0.072$, $d = 0.611$, $\lambda = 0.022$, $\tau = 0.005$, $\beta \in [0,1]$, $\psi \in [0,1]$ and $\alpha \in [0.01,1]$. We consider the initial conditions as $s_h(0) = d_1 = 40$, $i_h(0) = d_2 = 2$, $r_h(0) = d_3 = 0$, $s_m(0) = d_4 = 500$, and $i_m(0) = c_5 = 10$. Figure 3.1a–c yields the consequences of order of the AB derivative on the human population. In Figure 3.2a,b, we show the impact of order of the AB derivative on several categories of mosquitoes. In Figure 3.3a–c, we demonstrate the impact of anti-malarial drugs on distinct groups of the human population. In Figure 3.4a,b, we show the

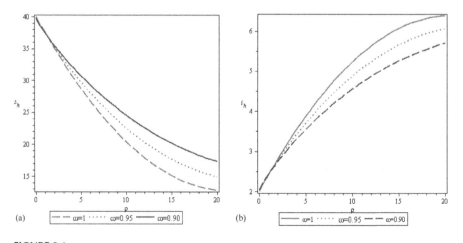

FIGURE 3.1
Impact of order of the AB derivative on the human population when $\psi = 0.05$, $\alpha = 0.05$ and $\beta = 0.05$: (a)s_h, (b)i_h. *(Continued)*

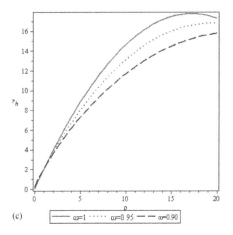

FIGURE 3.1 (Continued)
Impact of order of the AB derivative on the human population when $\psi = 0.05$, $\alpha = 0.05$ and $\beta = 0.05 : (c) r_h$.

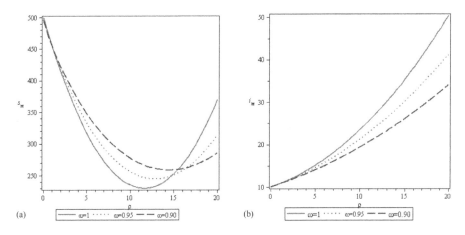

FIGURE 3.2
Impact of order of the AB derivative of an arbitrary order on the mosquito population when $\psi = 0.05$, $\alpha = 0.25$ and $\beta = 0.05 : (a) s_m, (b) i_m$.

effects of anti-malarial drugs on several categories of the mosquito population. In Figure 3.5a–c, we analyse the impact of the vaccination treatment on several categories of the human population. Lastly, in Figure 3.6a,b, we show the effect of the treatment of spraying on several groups of the mosquito population.

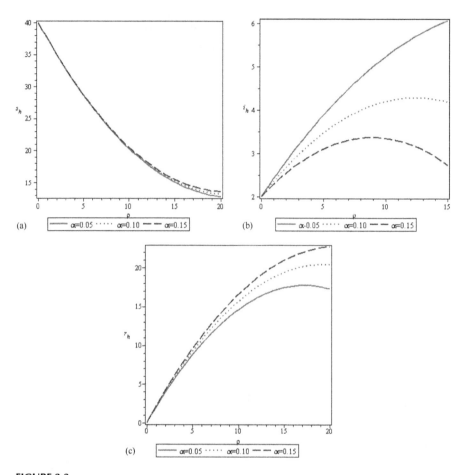

FIGURE 3.3
Effect of anti-malarial drugs on the human population when $\omega = 1$, $\psi = 0.05$ and $\beta = 0.05$: (a)s_h, (b)i_h and (c)r_h.

Fractional Calculus Approach in SIRS-SI Model

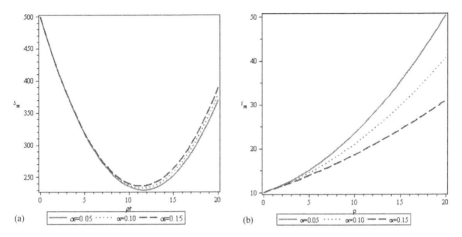

FIGURE 3.4
Impact of anti-malarial drugs on the mosquito population when $\omega = 1$, $\psi = 0.05$ and $\beta = 0.05 : (a) s_m, (b) i_m$.

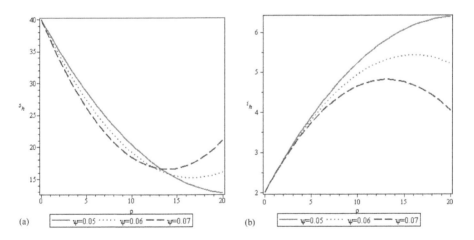

FIGURE 3.5
Influence of treatment of vaccines on the human population when $\omega = 1$, $\alpha = 0.05$ and $\beta = 0.05 : (a) s_h, (b) i_h$. *(Continued)*

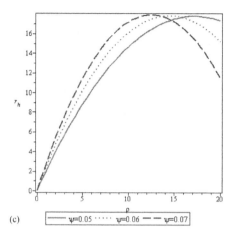

FIGURE 3.5 (Continued)
Influence of treatment of vaccines on the human population when $\omega = 1$, $\alpha = 0.05$ and $\beta = 0.05 : (c)r_h$.

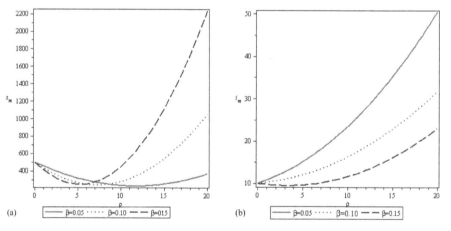

FIGURE 3.6
Effect of treatment of spraying on the mosquito population when $\omega = 1$, $\alpha = 0.25$ and $\psi = 0.05 : (a)s_h, (b)i_h$.

3.7 Conclusions

In this work, we discussed a SIRS-SI malaria disease model for describing the transmission of malaria disease involving a fractional-order derivative together with a number of cures. The main advantage of the discussed model lies in its strong memory effects. To analyse the EU of the solution of the model (Equation 3.5), the fixed-point postulate was applied. The newly introduced technique q-HATM was utilised for obtaining the numerical results. The effect of the order of the AB arbitrary order derivative on several classes of human and mosquito populations were studied. After observing the theoretical and numerical outcomes, we came to an end that the AB derivative is very important to represent the treatment and to manage the malaria disease and other kinds of problems.

Acknowledgements

This work is supported by the Competitive Research Scheme (CRS) project funded by the TEQIP-III (ATU) Rajasthan Technical University, Kota under Grant No. TEQIP-III/RTU(ATU)/CRS/2019-20/47.

References

1. M.B. Abdullahi, Y.A. Hasan, and F.A. Abdullah, A mathematical model of malaria and the effectiveness of drugs, *Appl. Math. Sci.*, 7 (62), 3079–3095 (2013).
2. S.I. Onah, O.C. Collins, C. Okoye, and G.C.E. Mbah, Dynamics and control measures for malaria using a mathematical epidemiological model, *J. Math. Anal. Appl.*, 7 (1), 65–73 (2019).
3. H.F. Huo and G.M. Qiu, Stability of a mathematical model of malaria transmission with relapse, *Abstr. Appl. Anal.*, Volume 2014, Article ID 289349, 9 pages, http://dx.doi.org/10.1155/2014/289349.
4. F.B. Agusto, N. Marcus, and K.O. Okosun, Application of optimal control to the epidemiology of malaria, *Elect. J. Differ. Eq.*, 2012 (81), 1–22 (2012).
5. C. Chiyaka, J.M. Tchuenche, W. Garira, and S. Dube, A mathematical analysis of the effects of control strategies on the transmission dynamics of malaria, *Appl. Math. Comput.*, 195, 641–662 (2008).
6. S. Mandal, R.R. Sarkar, and S. Sinha, Mathematical models of malaria: A review, *Malaria J.*, 10, 1–19 (2011).
7. M. Rafikov, L. Bevilacqua, and A.P.P. Wyse, Optimal control strategy of malaria vector using genetically modified mosquitoes, *J. Theor. Biol.*, 258, 418–425 (2009).
8. H.M. Yang, A mathematical model for malaria transmission relating global warming and local socioeconomic conditions, *Rev. Saude Publica*, 35 (3), 224–231 (2001).
9. R. Senthamarai, S. Balamuralitharan, and A. Govindarajan, Application of homotopy analysis method in SIRS-SI model of malaria disease, *Int. J. Pure Appl. Math.*, 113 (12), 239–248 (2017).
10. D. Kumar, J. Singh, M.A. Qurashi, and D. Baleanu, A new fractional SIRS-SI malaria disease model with application of vaccines, anti-malarial drugs, and spraying, *Adv. Differ. Equ.*, 2019, 278, (2019), https://doi.org/10.1186/s13662-019-2199-9.

11. M. Caputo, *Elasticita e dissipazione* (Zani-Chelli, Bologna, 1969).
12. A.A. Kilbas, H.M. Srivastava, and J.J. Trujillo, *Theory and Applications of Fractional Differential Equations* (Elsevier, Amsterdam, the Netherlands, 2006).
13. D. Baleanu, Z.B. Guvenc, and Z.A.T. Machado, *New Trends in Nanotechnology and Fractional Calculus Applications* (Springer, Dordrecht, Germany, 2010).
14. X.J. Yang, Z.A.T. Machado, D. Baleanu, and C. Cattani, On exact traveling-wave solutions for local fractional Korteweg-de Vries equation, *Chaos*, 26, 084312 (2016).
15. A. Atangana and R.T. Alqahtani, New numerical method and application to Keller-Segel model with fractional order derivative, *Chaos Solitons Fract.*, 116, 14–21 (2018).
16. J. Singh, D. Kumar, and J.J. Nieto, Analysis of an El Nino-Southern Oscillation model with a new fractional derivative, *Chaos Solitons Fract.*, 99, 109–115 (2017).
17. D. Kumar, J. Singh, and D. Baleanu, A new analysis of Fornberg-Whitham equation pertaining to a fractional derivative with Mittag-Leffler type kernel, *Eur. J. Phys.*, 133 (2), 70 (2018).
18. J. Singh, D. Kumar, D. Baleanu, and S. Rathore, An efficient numerical algorithm for the fractional Drinfeld-Sokolov-Wilson equation, *Appl. Math. Comput.*, 335, 12–24 (2018).
19. J. Singh, D. Kumar, and D. Baleanu, On the analysis of fractional diabetes model with exponential law, *Adv. Differ. Eq.*, 2018, 231 (2018).
20. Carla M.A. Pinto and Ana R.M. Carvalho, The role of synaptic transmission in a HIV model with memory, *Appl. Math. Comput.*, 292, 76–95 (2017).
21. Carla M.A. Pinto, Persistence of low levels of plasma viremia and of the latent reservoir in patients under ART: A fractional-order approach, *Comm. Nonlinear Sci. Numer. Simulat.*, 43, 251–260 (2017).
22. Carla M.A. Pinto and Ana R.M. Carvalho, Fractional complex-order model for HIV infection with drug resistance during therapy, *J. Vib. Control* (2015), doi:10.1177/1077546315574964.
23. R.L. Magin, *Fractional Calculus in Bioengineering* (Begell House Inc. Publishers, Danbury, CT, 2006).
24. A. Atangana and D. Baleanu, New fractional derivative with nonlocal and non-singular kernel, theory and application to heat transfer model, *Therm. Sci.*, 20 (2), 763–769 (2016).
25. D. Kumar, J. Singh, and D. Baleanu, On the analysis of vibration equation involving a fractional derivative with Mittag-Leffler law, *Math. Methods Appl. Sci.*, 43 (1), 443–457 (2019).
26. D. Kumar, J. Singh, and D. Baleanu, Analysis of regularized long-wave equation associated with a new fractional operator with Mittag-Leffler type kernel, *Physica A*, 492, 155–167 (2018).
27. J. Singh, D. Kumar, and D. Baleanu, New aspects of fractional Biswas-Milovic model with Mittag-Leffler law, *Math. Model. Nat. Pheno.*, 14, 303 (2019).
28. J. Singh, A new analysis for fractional rumor spreading dynamical model in a social network with Mittag-Leffler law, *Chaos*, 29, 013137 (2019).
29. D. Kumar, J. Singh, and D. Baleanu, A new analysis for fractional model of regularized long-wave equation arising in ion acoustic plasma waves, *Math. Methods Appl. Sci.*, 40 (15), 5642–5653, (2017).
30. H.M. Srivastava, D. Kumar, and J. Singh, An efficient analytical technique for fractional model of vibration equation, *Appl. Math. Model.*, 45, 192–204 (2017).
31. M.A. El-Tawil and S.N. Huseen, The q-homotopy analysis method (q-HAM), *Int. J. Appl. Math. Mech.*, 8, 51–75 (2012).

32. M.A. El-Tawil and S.N. Huseen, On convergence of the q-homotopy analysis method, *Int. J. Contemp. Math. Sci.*, 8, 481–497 (2013).
33. S.A. Khuri, A Laplace decomposition algorithm applied to a class of nonlinear differential equations, *J. Appl. Math.*, 1, 141–155 (2001).
34. D. Kumar, J. Singh, and D. Baleanu, A fractional model of convective radial fins with temperature-dependent thermal conductivity, *Rom. Rep. Phys.*, 69(1), 103 (2017).
35. D. Kumar, R.P. Agarwal, and J. Singh, A modified numerical scheme and convergence analysis for fractional model of Lienard's equation, *J. Comput. Appl. Math.* (2017), https://doi.org/10.1016/j.cam.2017.03.011.
36. D. Kumar, J. Singh, and D. Baleanu, Analytic study of Allen-Cahn equation of fractional order, *Bull. Math. Anal. Appl.*, 1, 31–40 (2016).

4

Mathematical Modelling and Analysis of Fractional Epidemic Models Using Derivative with Exponential Kernel

Kolade M. Owolabi and Abdon Atangana

CONTENTS
4.1 Introduction 109
4.2 Numerical Method of Approximation 112
4.3 Main Models and Mathematical Analysis 114
    4.3.1 Linear Stability Analysis for the Temporal System 115
    4.3.2 Existence and Uniqueness of Solutions 117
4.4 Simulation Results 122
4.5 Conclusion 126
References 127

4.1 Introduction

The transmission of various infectious diseases has been well described with the concept of mathematical modelling, which has played an enormous role in describing the method of disease spread and control from one organism to another. A lot of research work has been done based on the dynamic behaviour of multicomponent epidemic models. In the present work, we study an epidemic system with delay that describes the temporal interaction of the susceptible class, infected population and the recovery class that interact in a non-linear fashion. Some reasonable mathematical models have been suggested to prevent the spread of diseases like Ebola, HIV and AIDS, measles, Hepatitis B virus diseases, malaria and a host of others [5,27,28].

In another development, recent research works have pointed out that modelling of real-life phenomena with a derivative of fractional order is most appropriate and correct in terms of accuracy and reliability [6,7,19–24,29]. Fractional calculus, which is the generalisation of standard integral and derivative to fractional cases, is known to have played an important role in various fields of engineering, finance and science. In recent decades, scientists and researchers have deeply examined this special field in various forms, see for example [2,4,8,9,14,26].

In the present work, the fractional delayed epidemic system with a non-linear incidence rate that depends on the ratio of the numbers of susceptible and infectious population is considered. In such models, classical time derivatives are replaced with the Caputo-Fabrizio fractional order case. There are several existing mathematical definitions of fractional derivatives, but we only report here the cases of Caputo and Fabrizio derivatives that are paramount to our study.

Definition 4.1

The Riemann-Liouville and the Caputo fractional derivatives are defined as [2,8,13,17,18,24,25]

$$^{RL}D_{0,t}^{\alpha}u(t) = \frac{1}{\Gamma(n-\alpha)}\frac{d^n}{dt^n}\int_0^t (t-s)^{n-\alpha-1}u(s)ds \qquad (4.1)$$

and

$$^{C}D_{0,t}^{\alpha}u(t) = \frac{1}{\Gamma(n-\alpha)}\int_0^t (t-s)^{n-\alpha-1}\frac{d^n}{dt^n}u(s)ds \qquad (4.2)$$

respectively, with $n-1 < \alpha \leq n$. The formulation of other new fractional derivative definitions are based on either of these two operators.

Definition 4.2

If $y(t) \in H^1(a,b)$, for $b > a$, then the Caputo-Fabrizio derivative with fractional order $\gamma \in (0,1]$ is defined as [3]

$$D_t^{\gamma}y(t) = \frac{M(\gamma)}{1-\gamma}\int_a^t y'(s)\exp\left[-\gamma\frac{t-s}{1-\gamma}\right]ds \qquad (4.3)$$

where $M(\gamma)$ is the usual normalised function satisfying the condition $M(0) = 1, M(1) = 1$, see [3]. If $y(t) \notin H^1(a,b)$ then, the Caputo-Fabrizio operator is written as

$$D_t^{\gamma}y(t) = \frac{\gamma M(\gamma)}{1-\gamma}\int_a^t (y(t)-y(s))\exp\left[-\gamma\frac{t-s}{1-\gamma}\right]ds. \qquad (4.4)$$

It should be noted that with transformations $\xi = \frac{1-\gamma}{\gamma} \in [0,\infty)$ and $\gamma = \frac{1}{1+\xi} \in (0,1]$, the preceding equation becomes

$$D_t^\gamma y(t) = \frac{G(\xi)}{\xi} \int_a^t y'(t) \exp\left[-\frac{t-s}{\xi}\right] ds, G(0) = 1, G(\infty) = 1. \quad (4.5)$$

Also, $\lim_{\xi \to 0} \frac{1}{\xi} \exp\left[-\frac{t-s}{\xi}\right] = \delta(s-t)$.

Definition 4.3

Let $\gamma \in (0,1)$, then the fractional integral for a function $y(t)$ of order γ is defined as

$$I_t^\gamma y(t) = \frac{2(1-\gamma)}{M(\gamma)(2-\gamma)} y(t) + \frac{2\gamma}{M(\gamma)(2-\gamma)} \int_0^t y(s) ds, t \geq 0, \quad (4.6)$$

it requires from the above that

$$\frac{2(1-\gamma)}{M(\gamma)(2-\gamma)} + \frac{2\gamma}{M(\gamma)(2-\gamma)} = 1$$

which means $M(\gamma) = \frac{2}{2-\gamma}$ for $\gamma \in (0,1)$. As a result, the authors in [12] formulate a new version of the Caputo-Fabrizio operator given as

$$D_t^\gamma y(t) = \frac{1}{1-\gamma} \int_a^t y'(s) \exp\left[-\gamma \frac{t-s}{1-\gamma}\right] ds. \quad (4.7)$$

This derivative has been applied in number of ways to model real-life precesses. For example, Singh et al. [26] studied an epidemiological model of computer viruses in the sense of the Caputo-Fabrizio fractional derivative. Also, some numerical approximation techniques for the Caputo-Fabrizio derivative have been suggested. Owolabi and Atangana [15] considered methods of approximation of the spatial fractional derivative with the exponential decay law by applying the Caputo-Fabrizio derivative of order $0 < \gamma \leq 2$ in the Riemann-Liouville sense to study the superdiffusive process of parabolic differential equations.

The remainder part of this chapter is outlined as follows. In Section 4.2, we suggest a numerical method based on the Adams-Bashforth scheme for the approximation of the Caputo-Fabrizio operator. The disease-free equilibrium point is analysed for both local and global stability analysis in Section 4.3, and we also verify the existence and uniqueness of solutions. A range of numerical results obtained for different instances of fractional order are given in Section 4.4 to validate our theoretical findings. We finally conclude with the last section.

4.2 Numerical Method of Approximation

In this section, we adapt closely the three-step Adams-Bashforth method developed by Owolabi and Atangana [16] to numerically approximate the Caputo-Fabrizio operator whose formulation is based on the exponential kernel.

Given the Caputo-Fabrizio differential problem,

$$D_t^\gamma y(t) = f(t, y(t)) \tag{4.8}$$

or

$$\frac{M(\gamma)}{1-\gamma} \int_0^t y'(s) \exp\left[-\frac{\gamma}{1-\gamma}(t-s)\right] ds = f(t, y(t)). \tag{4.9}$$

By applying the fundamental theorem of calculus, the preceding equation transforms into

$$y(t) - y(0) = \frac{1-\gamma}{M(\gamma)} f(t, y(t)) + \frac{\gamma}{M(\gamma)} \int_0^t f(s, y(s)) ds \tag{4.10}$$

so that

$$y(t_{n+1}) - y(0) = \frac{1-\gamma}{M(\gamma)} f(t_n, y(t_n)) + \frac{\gamma}{M(\gamma)} \int_0^{t_{n+1}} f(t, y(t)) dt \tag{4.11}$$

and

$$y(t_n) - y(0) = \frac{1-\gamma}{M(\gamma)} f(t_{n-1}, y(t_{n-1})) + \frac{\gamma}{M(\gamma)} \int_0^{t_n} f(t, y(t)) dt \tag{4.12}$$

On subtracting equation (4.12) from equation (4.11), we have

$$y(t_{n+1}) - y(t_n) = \frac{1-\gamma}{M(\gamma)} \{f(t_n, y_n) - f(t_{n-1}, y_{n-1})\} + \frac{\gamma}{M(\gamma)} \int_{t_n}^{t_{n+1}} f(t, y(t)) dt \tag{4.13}$$

where

$$\int_{t_n}^{t_{n+1}} f(t, y(t)) dt = \int_{t_n}^{t_{n+1}} \left\{ \frac{f(t_n, y_n)}{h}(t-t_n) - \frac{f(t_{n-1}, y_{n-1})}{h}(t-t_{n-1}) \right.$$

$$\left. + \frac{f(t_{n-2}, y_{n-2})}{h}(t-t_n) \right\} dt \tag{4.14}$$

$$= \frac{23h}{12} f(t_n, y_n) - \frac{16h}{12} f(t_{n-1}, y_{n-1}) + \frac{5h}{12} f(t_{n-2}, y_{n-2}).$$

Thus,

$$y(t_{n+1}) - y(t_n) = \frac{1-\gamma}{M(\gamma)}\left[f(t_n, y_n) - f(t_{n-1}, y_{n-1})\right]$$

$$+ \frac{\gamma h}{12 M(\gamma)}\left[23 f(t_n, y_n) - 16 f(t_{n-1}, y_{n-1}) + 5 f(t_{n-2}, y_{n-2})\right]$$

which means

$$y(t_{n+1}) - y(t_n) = \left(\frac{1-\gamma}{M(\gamma)} + \frac{23\gamma h}{12 M(\gamma)}\right) f(t_n, y_n) - \left(\frac{1-\gamma}{M(\gamma)} + \frac{16\gamma h}{12 M(\gamma)}\right) f(t_{n-1}, y_{n-1})$$

$$+ \frac{5\gamma h}{12 M(\gamma)} f(t_{n-2}, y_{n-2}). \tag{4.15}$$

Hence,

$$y_{n+1} = y_n + \left(\frac{1-\gamma}{M(\gamma)} + \frac{23\gamma h}{12 M(\gamma)}\right) f(t_n, y_n) - \left(\frac{1-\gamma}{M(\gamma)} + \frac{16\gamma h}{12 M(\gamma)}\right) f(t_{n-1}, y_{n-1})$$

$$+ \frac{5\gamma h}{12 M(\gamma)} f(t_{n-2}, y_{n-2}) + R_n^{\gamma}(t) \tag{4.16}$$

which is called the three-step Adams-Bashforth algorithm [16] for the numerical approximation of the Caputo-Fabrizio derivative, where

$$R_n^{\gamma}(t) = \frac{\gamma}{M(\gamma)} \int_0^t \frac{3}{8} f^{(4)}(s) h^3 ds$$

$$\| R_n^{\gamma}(t) \|_{\infty} = \frac{\gamma}{M(\gamma)} \left\| \int_0^t \frac{3}{8} f^{(4)}(s) h^3 \right\|_{\infty} ds$$

$$\leq \frac{3\gamma h^3}{8 M(\gamma)} \int_0^t \left\| f^{(4)}(s) \right\|_{\infty} ds \tag{4.17}$$

$$\leq \frac{3\gamma h^3}{8 M(\gamma)} \cdot T_{\max}(\beta).$$

where

$$\beta = \max_{s \in [0,t]} \left\| f^{(4)}(s) \right\|_{\infty}.$$

Obviously, when $\gamma = 1$ in equation (4.16), we get the standard Adams-Bashforth three-step method

$$y_{n+1} = y_n + \frac{h[23f(t_n, y_n) - 16f(t_{n-1}, y_{n-1}) + 5f(t_{n-2}, y_{n-2})]}{12}. \qquad (4.18)$$

4.3 Main Models and Mathematical Analysis

In this segment, we introduce the main epidemic model and consider its various analysis. We let $u(t)$ be the number of susceptible individuals, and $v(t)$ and $w(t)$ stand for the infected and recovered population classes at time t, respectively. The interaction of these groups of individuals is represented in a non-linear fashion as

$$\frac{du}{dt} = f_1(u, v, w) = \sigma(u(t) - u^2(t)) - \frac{u(t)(t-\mu)v(t-\mu)}{1 + \alpha(t-\mu)v(t)}$$

$$\frac{dv}{dt} = f_2(u, v, w) = \frac{u(t)(t-\mu)v(t)(t-\mu)}{1 + \alpha(t-\mu)v(t)} - (\phi + \varphi + \psi)v(t), \qquad (4.19)$$

$$\frac{dw}{dt} = f_3(u, v, w) = \psi v(t) - \phi w(t)$$

where $f_i(u, v, w)$, for $i = 1, 2, 3$ represent the local kinetic reactions expected to be linear or nonlinear, σ is the intrinsic birth rate for the susceptible class, ϕ is the natural death rate, α stands for the inhibitory or psychological effect, and μ measures the latent period. The death rate due to disease is measured by φ, and ψ denotes the recovery rate. The response term $u(t)(t-\mu)v(t-\mu)/[1+\alpha(t-\mu)v(t)]$ is considered as the incidence rate. It is assumed that all the parameters are non-negative.

Recent developments have shown that modelling of real-life phenomena with non-integer order derivative is most accurate and reliable [24,26]. Hence, we are motivated in this work by replacing the ordinary or classical time derivative with the Caputo-Fabrizio fractional derivative of order $\gamma \in (0,1]$ as

$$D_t^\gamma u(t) = f_1(u, v, w) = \sigma(u(t) - u^2(t)) - \frac{u(t)(t-\mu)v(t-\mu)}{1 + \alpha(t-\mu)v(t)}$$

$$D_t^\gamma v(t) = f_2(u, v, w) = \frac{u(t)(t-\mu)v(t)(t-\mu)}{1 + \alpha(t-\mu)v(t)} - (\phi + \varphi + \psi)v(t), \qquad (4.20)$$

$$D_t^\gamma w(t) = f_3(u, v, w) = \psi v(t) - \phi w(t)$$

subject to the initial conditions

$$u(0) = \vartheta_1,\ v(0) = \vartheta_2,\ w(0) = \vartheta_3 \qquad (4.21)$$

where $(\vartheta_1, \vartheta_2, \vartheta_3) \in R_+ = R([-\mu, 0], R_+^3)$, for $\vartheta_i \geq 0, i = 1, 2, 3$, and we define $R_+^3 = \{\xi_1, \xi_2, \xi_3 \in R^3 : \xi_i \geq 0, i = 1, 2, 3\}$. For initial condition (4.21), the feasible region is given as

$$\Omega = \{(u, v, w) \in R_+^3 \mid u + v + w \leq 1\}.$$

Also, we take the total population $P = u + v + w$, assumed to be constant. We assume $\mathcal{B}$ to be the Banach space and $\mathbb{R} \to \mathbb{R}$ be continuous valued function on interval L with $\|(u, v, w)\| = \|u\| + \|v\| + \|w\|$, where $\|u\| = \sup\{|u(t)| : t \in L\}$, $\|v\| = \sup\{|v(t)| : t \in L\}$ and $\|u\| = \sup\{|u(t)| : t \in L\}$. In other words, $\mathcal{B} = \mathcal{B}(L) \times \mathcal{B}(L)$ with $\mathcal{B}(L)$ denotes the Banach space defined on L.

4.3.1 Linear Stability Analysis for the Temporal System

By setting $f_1(u, v, w) = 0, f_2(u, v, w) = 0$ and $f_2(u, v, w) = 0$, a simple calculation shows that the epidemic system seen in the previous section has a disease-free equilibrium state $E_0(1, 0, 0)$, which corresponds to the existence of population class of u only. We define the basic reproduction number as $R_0 = 1/q$, where $q = \phi + \varphi + \psi$. If $R_0 > 1$, we obtain a non-trivial state that corresponds to the existence of susceptible, infected and recovered individuals. This equilibrium state is denoted by

$$E^* \left\{ \frac{\sigma\alpha - 1 + \sqrt{(1 - \sigma\alpha)^2 + 4\sigma q \alpha}}{2\sigma\alpha},\ \frac{u^* - q}{\alpha q},\ \frac{(u^* - q)}{\phi \alpha q} \right\}.$$

In what follows, we show that the disease-free state is globally asymptotically stable while the endemic state is locally asymptotically stable for $0 \leq \mu \leq \mu_0$, and unstable if $\mu > \mu_0$.

Theorem 4.1

The disease-free equilibrium state E_0 is locally asymptotically stable if the reproduction number $R_0 < 1$ for any μ. It is unstable if otherwise.

Proof

The characteristic equation for $\mu \neq 0$ for the epidemic system is given as

$$(\lambda^2 + (\phi + \sigma)\lambda + \sigma\phi)(\lambda + q - e^{-\lambda\mu}) = 0 \qquad (4.22)$$

with negative roots $\lambda_1 = -\sigma$ and $\lambda_2 = -\phi$, others are expressed by the equation

$$\lambda + q - e^{-\lambda\mu} = 0, \lambda = e^{-\lambda\mu} - q, R_e(\lambda) = e^{-R_e(\lambda)}\cos(\mu kx\lambda) - q. \quad (4.23)$$

If $R_e(\lambda) \geq 0$, then $R_e(\lambda) \leq 1-q$ and $R_e(\lambda) \leq (R_0 - 1)q$. Since $R_0 < 1$, we have $R_e(\lambda) < 0$ Hence, the steady state E_0 for the temporal epidemic system is locally asymptotically stable. But if the reproduction number R_0 is greater than unity, for instance, if we let $g(\lambda) = \lambda + q - e^{-\lambda\mu}$, due to the fact that $g(0) < 0$ and $g(\infty) > 0$, we obtain positive real roots. Hence, the system at E_0 becomes unstable. □

Theorem 4.2

The disease-free equilibrium point E_0 for the epidemic system is globally asymptotically stable, if the basic reproduction number R_0 is less than unity for all values of μ.

Proof

We follow the stability idea in [10,11] and define a differentiable (Lyapunov) function

$$\mathcal{F} = F_1 + F_2, \quad (4.24)$$

when $F \geq 0$, we have

$$F_1 = I,$$

$$F_2 = \int_{t-\mu}^{t} \frac{u(t-\mu)}{1 + \alpha I(t-\mu)} d\zeta. \quad (4.25)$$

By finding the derivative of $\mathcal{F}(t)$, we obtain

$$\mathcal{F}' = F_1' + F_2',$$

$$\mathcal{F}' = \frac{uI}{1+\alpha I} - qI \quad (4.26)$$

since $u \leq 1$, then $\frac{1}{1+\alpha I} < 1$. So, when $R_0 < 1$, the derivative $\mathcal{F}' \leq (1-q)I$ in such that

$$\mathcal{F}' \leq (R_0 - 1)qI, \mathcal{F}' \leq 0.$$

From above, $\mathcal{F}' = 0$ provided $R = 0, I = 0$ and $u = 0$. At any time t, it can be verified that the point E_0 is a subset of $\{(u,v,w): \mathcal{F}' = 0\}$. Hence, the point E_0 of the epidemic system is globally asymptotically stable. □

4.3.2 Existence and Uniqueness of Solutions

By applying fractional integral defined in equations (4.6) to (4.20), we get

$$u(t) - u_0(t) = \frac{2(1-\gamma)}{M(\gamma)(2-\gamma)}\left[\sigma(u-u^2) - \frac{u(t-\mu)v(t-\mu)}{1+\alpha(t-\mu)v}\right]$$

$$+ \frac{2\gamma}{M(\gamma)(2-\gamma)}\int_0^t\left[\sigma(u-u^2) - \frac{u(s-\mu)v(s-\mu)}{1+\alpha(s-\mu)v}\right]ds$$

$$v(t) - v_0(t) = \frac{2(1-\gamma)}{M(\gamma)(2-\gamma)}\left[\frac{u(t-\mu)v(t-\mu)}{1+\alpha(t-\mu)v} - (\phi+\alpha_1+\psi)v\right]$$
(4.27)

$$+ \frac{2\gamma}{M(\gamma)(2-\gamma)}\int_0^t\left[\frac{u(s-\mu)v(s-\mu)}{1+\alpha(s-\mu)v} - (\phi+\alpha_1+\psi)v\right]ds$$

$$w(t) - w_0(t) = \frac{2(1-\gamma)}{M(\gamma)(2-\gamma)}[\psi v - \phi w] + \frac{2\gamma}{M(\gamma)(2-\gamma)}\int_0^t[\psi v - \phi w]ds$$

We follow the technique used in [1] to show the existence and uniqueness of solutions. For convenience, the kernels are represented by

$$K(t, u(t)) = \sigma(u(t) - u^2(t)) - \frac{u(t)(t-\mu)v(t-\mu)}{1+\alpha(t-\mu)v(t)}$$

$$K(t, v(t)) = \frac{u(t)(t-\mu)v(t)(t-\mu)}{1+\alpha(t-\mu)v(t)} - (\mu+\alpha_1+\psi)v(t)$$
(4.28)

$$K(t, w(t)) = \psi v(t) - \mu w(t).$$

We let $P : C \to C$ be a compact operator. So that,

$$Pu(t) = \frac{2(1-\gamma)}{M(\gamma)(2-\gamma)}K(t, u(t)) + \frac{2\gamma}{M(\gamma)(2-\gamma)}\int_0^t K(s, u(s))ds,$$

$$Pv(t) = \frac{2(1-\gamma)}{M(\gamma)(2-\gamma)}K(t, v(t)) + \frac{2\gamma}{M(\gamma)(2-\gamma)}\int_0^t K(s, v(s))ds \quad (4.29)$$

$$Pw(t) = \frac{2(1-\gamma)}{M(\gamma)(2-\gamma)}K(t, w(t)) + \frac{2\gamma}{M(\gamma)(2-\gamma)}\int_0^t K(s, w(s))ds$$

Theorem 4.3

The mapping $P : \mathcal{C} \to \mathcal{C}$ is continuous everywhere.

Proof

Assume $F \subset \mathcal{C}$ is bounded. There exists $\kappa_i > 0$, $i = 1, 2, 3$, in such that $\|u\| < \kappa_1$, $\|v\| < \kappa_2$ and $\|w\| < \kappa_3$. We let

$$G_1 = \max_{0 \le t \le 1, 0 \le u \le \kappa_1} K(t, u(t)), \quad G_2 = \max_{0 \le t \le 1, 0 \le v \le \kappa_1} K(t, v(t)) \quad \text{and} \quad G_3 = \max_{0 \le t \le 1, 0 \le w \le \kappa_1} K(t, w(t)).$$

For all $u, v, w \in F$, we have

$$|Pu(t)| = \left| \frac{2(1-\gamma)}{M(\gamma)(2-\gamma)} K(t, u(t)) + \frac{2\gamma}{M(\gamma)(2-\gamma)} \int_0^t K(s, u(s)) ds \right|,$$

$$\le \left| \frac{2(1-\gamma)}{M(\gamma)(2-\gamma)} \right| K(t, u(t)) + \left| \frac{2\gamma}{M(\gamma)(2-\gamma)} \right| \int_0^t K(s, u(s)) ds,$$

$$\le \left\{ \frac{2(1-\gamma)}{M(\gamma)(2-\gamma)} + \frac{2\gamma}{M(\gamma)(2-\gamma)} \rho \right\} |K(t, u(t))|, \qquad (4.30)$$

$$\le \left\{ \frac{2(1-\gamma)}{M(\gamma)(2-\gamma)} + \frac{2\gamma}{M(\gamma)(2-\gamma)} \rho \right\} |G_1|,$$

$$\le \frac{2G_1}{M(\gamma)(2-\gamma)} (1 - \gamma + \gamma\rho).$$

In the same manner,

$$|Pv(t)| = \left| \frac{2(1-\gamma)}{M(\gamma)(2-\gamma)} K(t, v(t)) + \frac{2\gamma}{M(\gamma)(2-\gamma)} \int_0^t K(s, v(s)) ds \right|,$$

$$\le \left| \frac{2(1-\gamma)}{M(\gamma)(2-\gamma)} \right| K(t, v(t)) + \left| \frac{2\gamma}{M(\gamma)(2-\gamma)} \right| \int_0^t K(s, v(s)) ds,$$

$$\le \left\{ \frac{2(1-\gamma)}{M(\gamma)(2-\gamma)} + \frac{2\gamma}{M(\gamma)(2-\gamma)} \rho \right\} |K(t, u(t))|, \qquad (4.31)$$

$$\le \left\{ \frac{2(1-\gamma)}{M(\gamma)(2-\gamma)} + \frac{2\gamma}{M(\gamma)(2-\gamma)} \rho \right\} |G_2|,$$

$$\le \frac{2G_2}{M(\gamma)(2-\gamma)} (1 - \gamma + \gamma\rho).$$

and

$$|Pw(t)| = \left| \frac{2(1-\gamma)}{M(\gamma)(2-\gamma)} K(t, w(t)) + \frac{2\gamma}{M(\gamma)(2-\gamma)} \int_0^t K(s, w(s))ds \right|,$$

$$\leq \left| \frac{2(1-\gamma)}{M(\gamma)(2-\gamma)} \right| |K(t, w(t))| + \left| \frac{2\gamma}{M(\gamma)(2-\gamma)} \right| \int_0^t K(s, w(s))ds,$$

$$\leq \left\{ \frac{2(1-\gamma)}{M(\gamma)(2-\gamma)} + \frac{2\gamma}{M(\gamma)(2-\gamma)} \eta \right\} |K(t, u(t))|, \qquad (4.32)$$

$$\leq \left\{ \frac{2(1-\gamma)}{M(\gamma)(2-\gamma)} + \frac{2\gamma}{M(\gamma)(2-\gamma)} \eta \right\} |G_3|,$$

$$\leq \frac{2G_3}{M(\gamma)(2-\gamma)} (1-\gamma+\gamma\eta).$$

Therefore, $P(F)$ is bounded. In what follows, we let $t_1 < t_2$ and $u, v, w \in F$, for any $\varepsilon > 0$, there exists $|t_2 - t_1| < \delta$. Now

$$\|Pu(t_2) - Pu(t_1)\| = \left| \frac{2(1-\gamma)}{M(\gamma)(2-\gamma)} [K(t_2, u(t_2)) - K(t_1, u(t_1))] \right.$$

$$\left. + \left| \frac{2\gamma}{M(\gamma)(2-\gamma)} \int_0^{t_2} K(s, u(s))ds - \frac{2\gamma}{M(\gamma)(2-\gamma)} \int_0^{t_1} K(s, u(s))ds \right| \right. \qquad (4.33)$$

$$\leq \frac{2(1-\gamma)}{M(\gamma)(2-\gamma)} |K(t_2, u(t_2)) - K(t_1, u(t_1))|$$

$$+ \frac{2\gamma}{M(\gamma)(2-\gamma)} G_1 |K(t_2, u(t_2)) - K(t_1, u(t_1))|.$$

Next, we check

$$|K(t_2, u(t_2)) - K(t_1, u(t_1))| \leq \left| \sigma[u(t_2) - u(t_1)](1 - (u(t_2) - u(t_1))) \right.$$

$$\left. - \frac{\beta^2(u(t_2) - u(t_1))v}{1 + \alpha\beta v} \right| \qquad (4.34)$$

$$\leq \theta_1 |u(t_2) - u(t_1)| - \theta_2 |u(t_2) - u(t_1)|$$

$$\leq (\theta_1 - \theta_2) |u(t_2) - u(t_1)|$$

$$\leq A |t_2 - t_1|.$$

By using the last two preceding equations, we obtain

$$|Pu(t_2) - Pu(t_1)| \leq \frac{2\gamma}{M(\gamma)(2-\gamma)} A |t_2 - t_1| + \frac{2(1-\gamma)}{M(\gamma)(2-\gamma)} G_1 |t_2 - t_1|, \quad (4.35)$$

and

$$\delta = \varepsilon / \frac{2\gamma A}{M(\gamma)(2-\gamma)} + \frac{2(1-\gamma)G_1}{M(\gamma)(2-\gamma)}, \quad (4.36)$$

Hence, $|Pu(t_2) - Pu(t_1)| \leq \varepsilon$. A similar argument holds for the components v and w to obtain

$$\delta = \varepsilon / \frac{2\gamma B}{M(\gamma)(2-\gamma)} + \frac{2(1-\gamma)G_2}{M(\gamma)(2-\gamma)}, \quad (4.37)$$

and

$$\delta = \varepsilon / \frac{2\gamma C}{M(\gamma)(2-\gamma)} + \frac{2(1-\gamma)G_3}{M(\gamma)(2-\gamma)}, \quad (4.38)$$

respectively. Also, $|Pv(t_2) - Pv(t_1)| \leq \varepsilon$ and $|Pw(t_2) - Pw(t_1)| \leq \varepsilon$ are satisfied. Hence, $P(F)$ is continuous, while $\overline{P(F)}$ is compact.

For the uniqueness of the solution, we have

$$|Pu_1(t) - Pu_2(t)| = \left| \frac{2(1-\gamma)}{M(\gamma)(2-\gamma)} [K(t, u_1(t)) - K(t, u_2(t))] \right.$$

$$+ \frac{2\gamma}{M(\gamma)(2-\gamma)} \int_0^t [K(s, u_1(s)) - K(s, u_2(s))] ds$$

$$= \frac{2(1-\gamma)}{M(\gamma)(2-\gamma)} |[K(t, u_1(t)) - K(t, u_2(t))]| \quad (4.39)$$

$$+ \frac{2\gamma}{M(\gamma)(2-\gamma)} \left| \int_0^t [K(s, u_1(s)) - K(s, u_2(s))] ds \right|$$

$$\leq \frac{2(1-\gamma)}{M(\gamma)(2-\gamma)} H_1 |u_1(t) - u_2(t)| + \frac{2\gamma}{M(\gamma)(2-\gamma)} H_1 |u_1(t) - u_2(t)|.$$

In a more compact form, the preceding equation becomes

$$|Pu_1(t) - Pu_2(t)| = \left\{ \frac{2(1-\gamma)H_1}{M(\gamma)(2-\gamma)} + \frac{2\gamma H_1}{M(\gamma)(2-\gamma)} \right\} |u_1(t) - u_2(t)|.$$

The preceding procedures are repeated for the remaining components to obtain

$$|Pv_1(t) - Pv_2(t)| = \left|\frac{2(1-\gamma)}{M(\gamma)(2-\gamma)}[K(t,v_1(t)) - K(t,v_2(t))]\right.$$

$$\left. + \frac{2\gamma}{M(\gamma)(2-\gamma)}\int_0^t [K(s,v_1(s)) - K(s,v_2(s))]ds\right|$$

$$= \frac{2(1-\gamma)}{M(\gamma)(2-\gamma)}|[K(t,v_1(t)) - K(t,v_2(t))]| \qquad (4.40)$$

$$+ \frac{2\gamma}{M(\gamma)(2-\gamma)}\left|\int_0^t [K(s,v_1(s)) - K(s,v_2(s))]ds\right|$$

$$\leq \frac{2(1-\gamma)}{M(\gamma)(2-\gamma)}H_v|v_1(t) - v_2(t)|$$

$$+ \frac{2\gamma}{M(\gamma)(2-\gamma)}H_v|v_1(t) - v_2(t)|.$$

which is written as

$$|Pv_1(t) - Pv_2(t)| = \left\{\frac{2(1-\gamma)H_2}{M(\gamma)(2-\gamma)} + \frac{2\gamma H_2}{M(\gamma)(2-\gamma)}\right\}|v_1(t) - v_2(t)|.$$

and

$$|Pw_1(t) - Pw_2(t)| = \left|\frac{2(1-\gamma)}{M(\gamma)(2-\gamma)}[K(t,w_1(t)) - K(t,w_2(t))]\right.$$

$$\left. + \frac{2\gamma}{M(\gamma)(2-\gamma)}\int_0^t [K(s,w_1(s)) - K(s,w_2(s))]ds\right| \qquad (4.41)$$

$$= \frac{2(1-\gamma)}{M(\gamma)(2-\gamma)}|[K(t,w_1(t)) - K(t,w_2(t))]|$$

$$+ \frac{2\gamma}{M(\gamma)(2-\gamma)}\left|\int_0^t [K(s,w_1(s)) - K(s,w_2(s))]ds\right|$$

$$\leq \frac{2(1-\gamma)}{M(\gamma)(2-\gamma)}H_3|w_1(t) - w_2(t)|$$

$$+ \frac{2\gamma}{M(\gamma)(2-\gamma)}H_3|w_1(t) - w_2(t)|.$$

which transforms into

$$|Pw_1(t) - Pw_2(t)| = \left\{\frac{2(1-\gamma)H_3}{M(\gamma)(2-\gamma)} + \frac{2\gamma H_3}{M(\gamma)(2-\gamma)}\right\}|w_1(t) - w_2(t)|.$$

Hence, if conditions

$$\left[\frac{2(1-\gamma)H_1}{M(\gamma)(2-\gamma)} + \frac{2\gamma H_1}{M(\gamma)(2-\gamma)}\right] < 1,$$

$$\left[\frac{2(1-\gamma)H_2}{M(\gamma)(2-\gamma)} + \frac{2\gamma H_2}{M(\gamma)(2-\gamma)}\right] < 1,$$

and

$$\left[\frac{2(1-\gamma)H_3}{M(\gamma)(2-\gamma)} + \frac{2\gamma H_3}{M(\gamma)(2-\gamma)}\right] < 1,$$

are satisfied, then P is a contraction, we conclude by saying that the fractional epidemic model defined in the sense of the Caputo-Fabrizio fractional derivative of order $0 < \gamma \leq 1$ has a unique solution.

4.4 Simulation Results

In the simulation experiments, we implement via the MATLAB package for the fractional three-step Adams-Bashforth method written compactly for the u component as

$$u_{n+1} = (1 + \chi_1)u_n - \chi_2 u_{n-1} + \chi_3 u_{n-2} \qquad (4.42)$$

where

$$\chi_1 = \left(\frac{1-\alpha}{M(\gamma)} + \frac{23\gamma h}{12M(\gamma)}\right),$$

$$\chi_2 = \left(\frac{1-\alpha}{M(\gamma)} + \frac{16\gamma h}{12M(\gamma)}\right), \qquad (4.43)$$

$$\chi_3 = \frac{5\gamma h}{12M(\gamma)},$$

and u_n is determined from the initial condition, and we predict the points u_{n-1} and u_{n-2} using MATLAB ode15s package.

We consider the epidemic model discussed in Section 4.2, that is

$$D_t^\gamma u(t) = \sigma(u(t) - u^2(t)) - \frac{u(t)(t-\mu)v(t-\mu)}{1+\alpha(t-\mu)v(t)} = f_1(u,v,w),$$

$$D_t^\gamma v(t) = \frac{u(t)(t-\mu)v(t)(t-\mu)}{1+\alpha(t-\mu)v(t)} - (\phi + \varphi + \psi)v(t) = f_2(u,v,w), \qquad (4.44)$$

$$D_t^\gamma w(t) = \psi v(t) - \phi w(t) = f_3(u,v,w)$$

with the initial conditions

$$u(0) = 0.5, \; v(0) = 0.5, \; w(0) = 0.5 \tag{4.45}$$

where $u(t), v(t), w(t)$ are the susceptible, infected and recovered individuals, respectively.

By using equation (4.42), the preceding equation is transformed into

$$\begin{aligned}u_{n+1} &= (1+\chi_1)f_1(t_n,u_n,v_n,w_n) - \chi_2 f_1(t_{n-1},u_{n-1},v_{n-1},w_{n-1}) + \chi_3 f_1(t_{n-2},u_{n-2},v_{n-2},w_{n-2})\\ v_{n+1} &= (1+\chi_1)f_2(t_n,u_n,v_n,w_n) - \chi_2 f_2(t_{n-1},u_{n-1},v_{n-1},w_{n-1}) + \chi_3 f_2(t_{n-2},u_{n-2},v_{n-2},w_{n-2})\\ w_{n+1} &= (1+\chi_1)f_3(t_n,u_n,v_n,w_n) - \chi_2 f_3(t_{n-1},u_{n-1},v_{n-1},w_{n-1}) + \chi_3 f_3(t_{n-2},u_{n-2},v_{n-2},w_{n-2})\end{aligned}$$

$$(4.46)$$

The behaviour of the dynamic system (equation 4.44) is shown in Figures 4.1 through 4.5 with parameters $\sigma = 0.0215$, $\mu = 2$, $\phi = 0.045$, $\varphi = 0.0162$, $\psi = 0.0324$, $\alpha = 0.013$. We can see in all cases that the endemic equilibrium point E^* is asymptotically stable over a period of time t for the fractional epidemic system. The results shown in Figure 4.4 correspond to a classical case when $\gamma = 1$. The presence of attractors show that the species coexist.

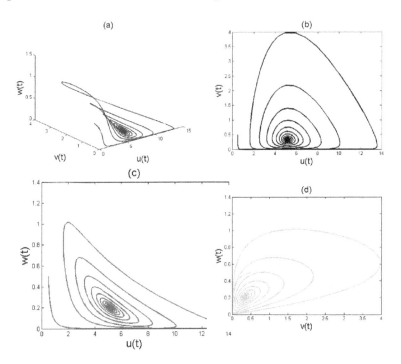

FIGURE 4.1
(a–d) Chaotic attractors in 2D and 3D for $\gamma = 0.5$.

124 *Fractional Calculus in Medical and Health Science*

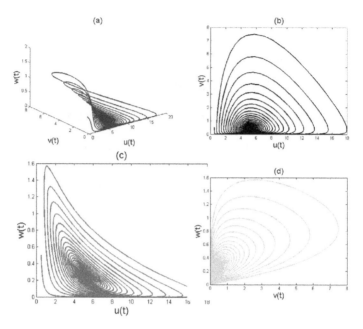

FIGURE 4.2
(a–d) Chaotic attractors in 2D and 3D for $\gamma = 0.65$.

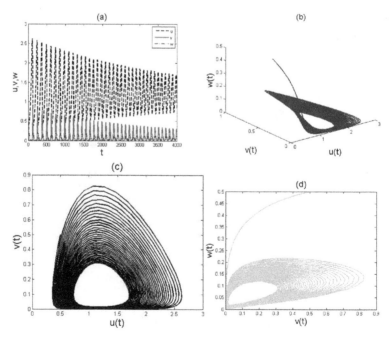

FIGURE 4.3
(a–d) Numerical results showing the distribution of species when $R_0 > 1$ with $\gamma = 0.88$.

Mathematical Modelling and Analysis of Fractional Epidemic Models

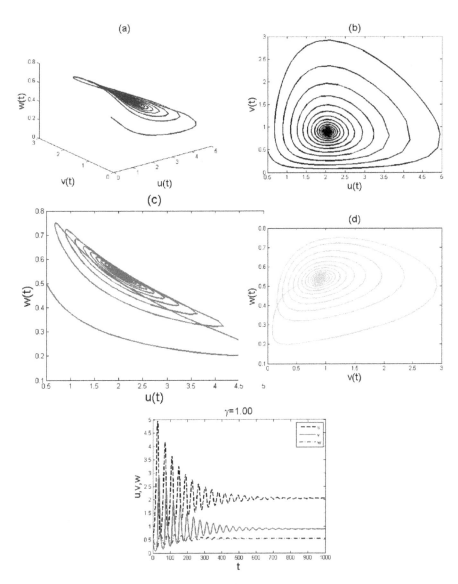

FIGURE 4.4
(a–d) Classical results for the epidemic system with $\gamma = 1.00$.

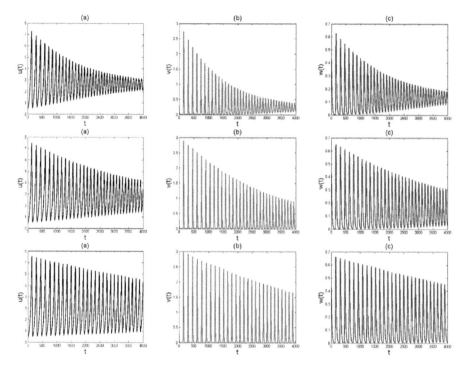

FIGURE 4.5
Numerical results for different instances of fractional values. The first, second and third rows correspond to $\gamma = (0.75, 0.82, 0.93)$, respectively.

4.5 Conclusion

In this chapter, the dynamics of the epidemic system is studied in the sense of the Caputo-Fabrizio fractional derivative. In the model, the classical time derivative is replaced with the fractional case. We prove the existence and uniqueness of solutions. In the system, we also show that the disease-free equilibrium point E_0 is globally asymptotically stable when the basic reproduction number R_0 is less than one for any delay μ, and the endemic equilibrium state is locally asymptotically stable when R_0 is greater than one with $\mu = 0$. A number of simulation results are presented for different values of γ to justify our theoretical findings. In the results, we observed that both species oscillate in phase regardless of the value of γ chosen. The presence of an attractor indicates the coexistence of the species.

References

1. A. Atangana and K.M. Owolabi, New numerical approach for fractional differential equations, *Mathematical Modelling of Natural Phenomena*, **13** (2018) 3.
2. M. Caputo, Linear models of dissipation whose Q is almost frequency independent II, *Geophysical Journal of the Royal Astronomical Society*, **13** (1967) 529–539.
3. M. Caputo and M. Fabrizio, A new definition of fractional derivative without singular kernel, *Progress in Fractional Differentiation and Applications*, **1** (2015) 73–85.
4. M.A. Dokuyucu, E. Celik, H. Bulut, and H.M. Baskonus, Cancer treatment model with the Caputo-Fabrizio fractional derivative, *The European Physical Journal Plus*, **133** (2018) 92.
5. Y. Enatsu, E. Messina, Y. Muroya, Y. Nakata, E. Russo, and A. Vecchio, Stability analysis of delayed SIR epidemic models with a class of nonlinear incidence rates, *Applied Mathematics and Computation*, **218** (2012) 5327–5336.
6. B. Karaagac, Analysis of the cable equation with non-local and non-singular kernel fractional derivative, *The European Physical Journal Plus*, **133** (2018) 54.
7. B. Karaagac, A study on fractional Klein Gordon equation with non-local and non-singular kernel, *Chaos, Solitons and Fractals*, **126** (2019) 218–229.
8. A.A. Kilbas, H.M. Srivastava, and J.J. Trujillo, *Theory and Applications of Fractional Differential Equations*, Elsevier, Amsterdam, the Netherlands, 2006.
9. D. Kumar, J. Singh, D. Baleanu, and M.A. Qurashi, Analysis of logistic equation pertaining to a new fractional derivative with non-singular kernel, *Advances in Mechanical Engineering*, **9** (2017) 1–8.
10. J.D. Murray, *Mathematical Biology I: An Introduction*, Springer-Verlag, New York, 2003.
11. J.D. Murray, *Mathematical Biology II: Spatial Models and Biomedical Applications*, Springer-Verlag, Berlin, Germany, 2003.
12. J. Losada and J.J. Nieto, Properties of the new fractional derivative without singular kernel, *Progress in Fractional Differentiation and Applications*, **1** (2015) 87–92.
13. K.B. Oldham and J. Spanier, *The Fractional Calculus: Theory and Applications of Differentiation and Integration to Arbitrary Order*, Dover Publications, New York, 2006.
14. K.M. Owolabi, Numerical solution of diffusive HBV model in a fractional medium, *SpringerPlus*, **5** (2016) 1643.
15. K.M. Owolabi and A. Atangana, Numerical approximation of nonlinear fractional parabolic differential equations with Caputo-Fabrizio derivative in Riemann-Liouville sense, *Chaos, Solitons and Fractals*, **99** (2017) 171–179.
16. K.M. Owolabi and A. Atangana, Analysis and application of new fractional Adams-Bashforth scheme with Caputo-Fabrizio derivative, *Chaos, Solitons and Fractals*, **105** (2017) 111–119.
17. K.M. Owolabi, Mathematical analysis and numerical simulation of chaotic noninteger order differential systems with Riemann-Liouville derivative, *Numerical Methods for Partial Differential Equations*, **34** (2018) 274–295.

18. K.M. Owolabi, Riemann-Liouville fractional derivative and application to model chaotic differential equations, *Progress in Fractional Differentiation and Applications*, **4** (2018) 99–110.
19. K.M. Owolabi, Analysis and numerical simulation of multicomponent system with Atangana-Baleanu fractional derivative, *Chaos, Solitons and Fractals*, **115** (2018) 127–134.
20. K.M. Owolabi, Numerical simulation of fractional-order reaction-diffusion equations with the Riesz and Caputo derivatives, *Neural Computing and Applications* (2019). doi:10.1007/s00521-019-04350-2.
21. K.M. Owolabi, Computational study of noninteger order system of predation, *Chaos*, **29** 013120 (2019).
22. K.M. Owolabi and Z. Hammouch, Mathematical modeling and analysis of two-variable system with noninteger-order derivative, *Chaos*, **29** (2019) 013145.
23. K.M. Owolabi and A. Atangana, On the formulation of Adams-Bashforth scheme with Atangana-Baleanu-Caputo fractional derivative to model chaotic problems, *Chaos*, **29** (2019) 023111.
24. I. Podlubny, *Fractional Differential Equations*, Academic Press, San Diego, CA, 1999.
25. S. Samko, A. Kilbas, and O. Marichev, *Fractional Integrals and Derivatives: Theory and Applications*, Gordon and Breach, Amsterdam, the Netherlands, 1993.
26. J. Singh, D. Kumar, Z. Hammouch, and A. Atangana, A fractional epidemiological model for computer viruses pertaining to a new fractional derivative, *Applied Mathematics and Computation*, **316** (2018) 504–515.
27. X. Song, X. Zhou, and X. Zhao, Properties of stability and Hopf bifurcation for a HIV infection model with time delay, *Applied Mathematical Modelling*, **34** (2010) 1511–1523.
28. L.S. Wen and X.F. Yang, Global stability of a delayed SIRS model with temporary immunity, *Chaos, Solitons and Fractals*, **38** (2008) 221–226.
29. S. Yadav, R.K. Pandey, and A.K. Shukla, Numerical approximations of Atangana-Baleanu Caputo derivative and its application, *Chaos, Solitons and Fractals*, **118** (2019) 58–64.

5
Fractional Order Mathematical Model for the Cell Cycle of a Tumour Cell

Ritu Agarwal, Kritika and Sunil Dutt Purohit

CONTENTS

5.1 Introduction 129
5.2 Mathematical Preliminaries 131
5.3 Mathematical Model of Cell Cycle for the Tumour Cell 132
5.4 Solution of the Model of the Cell Cycle for the Tumour Cell: Existence and Uniqueness 133
5.5 Result and Discussion 140
5.6 Conclusion 144
References 144

5.1 Introduction

Cancer is the most common severe disease in the world. Cancer refers to a disease characterised by the development of abnormal cells that divide uncontrollably and have the ability to infiltrate and destroy normal tissue. Certain forms of cancer result in a visible growth called a tumour. Nowadays, the mathematical model for the growth of a tumour has been developed and reviewed to better understand the disease mechanism and anticipate its future behaviour. Cancer is a disease that causes an uncontrolled division of cells. Among the several forms of cancer, this visible growth (by cell division) is called a tumour. So, in the analysis of tumour growth, the process of cell aging as well as its growth should be taken under consideration; [1–4] are several approaches in this direction. Mackey [5] presented a classical cell-cycle model: the "G_0" model. Webb [6] and Khiefetz et al. [7] developed models of cycle-specific chemotherapy. The process of cell aging and cell division cannot be neglected during the tumour growth investigation. There are two distinct processes through which cells can enter the cell cycle. These two processes are fertilisation and cell profileration activated by growth factors.

The cell cycle is a four-phase process among which the first one is the G_1 phase. This phase is basically the growth phase, and synthesis of various enzymes occurs in this phase that are needed for DNA replication in the next S phase. This phase is followed by the G_2 phase where significant biosynthesis occurs. The last phase of this cell cycle is of mitosis where division of the nucleus and cytoplasm results in two daughter cells, which again participate in the cell cycle by entering in the cycle via G_1. The first three phases, that is, G_1, S, and G_2 are noted as interphase. Here the primary focus will be on the cell cycle events that occur when G_0 cells are stimulated to proliferation by cell growth factors. The cell cycle duration for a normal cell is approximately 24 hours; the liver cell exceptionally takes a year to complete the cycle. Tubiana [8] studied 30 solid human tumours and suggested that the mean extent of the cell cycle is of two days shared as: one day alloted to G_1, 18 hours to S, 6 hours to G_2 and 1 hour to mitosis, M. The values mentioned here as per the study of Tubiana give the mean extent only, and various cell lines vary in cell cycle times (normal and cancerous). Over last few decades, there have been significant developments in theoretical, experimental and clinical approaches to understanding the dynamics of cancer cells and their interactions with the immune system. These have led to the development of important methods for cancer therapy, including vivotherapy, immunotherapy, chemotherapy, targeted drug therapy and many others. Along with this, there has also been some development on analytical and computational models to provide insights into clinical observations. In this work, we developed a mathematical model involving a system of governing differential equations that describes the dynamics of each of the interacting component cells. Specifically, the variables are the number of quiescent, interphase and mitotic cells with prescribed initial conditions.

For tumour growth, a number of models have been already studied, which are governed by ordinary differential equations. But the mathematical formulation of these models was lacking in non-locality effect and hence has declined to model the real-world problems. The concept of fractional calculus, which is a natural generalisation of classical calculus and the operators of differentiation, is a non-local operator. It is only in the past decade or so that fractional calculus has drawn the attention of main-stream science as a way to describe the dynamics of complex phenomena. Since, in modelling real-world problems [9–25] fractional operators play important roles. Bagley and Torvik [26] showed that the models of fractional order are more convenient than the integer order models. Fractional calculus provides a concise model for the description of the dynamic events that occur in biological tissues. Such a description is important for gaining an understanding of the underlying multiscale processes that occur. Many researchers have studied biological models possessing derivatives of arbitrary order and also found the numerical solution for them. Using a fractional advection diffusion equation, Agarwal et al. [27] studied the concentration profile of calcium. Arshad et al. [28] proposed a fractional model for human immunodeficiency virus (HIV) infection.

On the basis of the exponential decay law, differential operators with a non-singular kernel is recommended by virtue of Caputo and Fabrizio (C-F).

Fractional Order Mathematical Model for the Cell Cycle of a Tumour Cell

Losado and Nieto [29] provided the fundamental properties of this new operator. Several studies based upon this new operator were made [30–37]. The classical fractional differential operators among the singular kernel, especially the Caputo and Riemann, are unable to explain the memory effect precisely. Here the C-F fractional derivative is used to study the cell cycle of the tumour cell, since the kernel is used to explain the memory effect of the physical system.

5.2 Mathematical Preliminaries

The Caputo fractional derivative given through M. Caputo in 1967, has singularity in its kernel, and this issue was rectified in 2015 by Caputo and Fabrizio [38] with the introduction of a fresh operator having non-singular kernel. This new operator is lay it out as

Definition 5.1

The C-F fractional operator [38] of a function $h \in H^1(\alpha, \beta)$ (here, $H^1(\alpha, \beta)$, $\beta > \alpha$, is collection of functions that are differentiable sufficiently many times), of order $\eta \in [0,1)$ is represented as,

$$D_t^\eta (h(t)) = \frac{M(\eta)}{1-\eta} \int_\alpha^t h'(x) e^{-\eta \frac{t-x}{1-\eta}} dx, \quad \alpha < t < \beta. \tag{5.1}$$

Representation of the C-F differential operator in terms of convolution is,

$$D_t^\eta (h(t)) = \frac{M(\eta)}{1-\eta} h'(t) * e^{-\frac{\eta}{1-\eta} t}, \quad \eta \in [0,1]. \tag{5.2}$$

In the equation (5.1), $M(\eta)$ with the condition $M(0) = M(1) = 1$, represents the normalisation of the function. For $h \notin H^1(\alpha, \beta)$ the differentiation is defined as:

$$D_t^\eta (h(t)) = \frac{\eta M(\eta)}{1-\eta} \int_\eta^t (h(t) - h(x)) e^{-\eta \frac{t-x}{1-\eta}} dx. \tag{5.3}$$

Remark 5.1

If $\rho = \frac{1-\eta}{\eta} \in [0, \infty)$, $\eta = \frac{1}{1+\rho} \in [0,1]$, then equation (5.3) becomes

$$D_t^\eta(h(t)) = \frac{N(\rho)}{\rho}\int_\eta^t h'(\xi)\exp\left[-\frac{t-\xi}{\rho}\right]d\xi, \quad N(0) = N(\infty) = 1. \tag{5.4}$$

here $N(\rho)$ is normalisation term of $M(a)$. Moreover,

$$\lim_{\rho\to 0}\frac{1}{\rho}\exp\left[-\frac{t-\xi}{\rho}\right] = \delta(\xi - t). \tag{5.5}$$

Losada and Nieto [29] gave the relevant Caputo-Fabrizio integral.

Definition 5.2

The C-F integral of order a of the function h is given by

$$I_t^\eta(h(t)) = \frac{2(1-\eta)}{(2-\eta)M(\eta)}h(t) + \frac{2\eta}{(2-\eta)M(\eta)}\int_0^t h(s)ds, \quad t \geq 0. \tag{5.6}$$

Remark 5.2

On behalf of aforementioned definition, the C-F integral can be expressed in terms of average of a function h and its integral of order one.

$$\frac{2(1-\eta)}{(2-\eta)M(\eta)} + \frac{2\eta}{(2-\eta)(M(\eta))} = 1, \tag{5.7}$$

which implies $M(\eta) = \frac{2\eta}{2-\eta}$, $0 < \eta < 1$. From result of equation (5.7), Losado and Nieto defined the fractional derivative operator in the Caputo-Fabrizio sense of order a, $0 < \eta < 1$, as

$$D_t^\eta(h(t)) = \frac{1}{1-\eta}\int_\eta^t h'(\xi)\exp\left[\eta\frac{t-\xi}{1-\eta}\right]d\xi. \tag{5.8}$$

5.3 Mathematical Model of Cell Cycle for the Tumour Cell

In this section, we examine the cell cycle for the tumour cell given by [39]. Suppose the number of quiescent (G_0) cells at a time t is denoted by $Q(t)$. As the cell enters the cell cycle, it undergoes four phases, namely, G_1 (growth phase), S (synthesis phase), G_2 (second growth phase), and the concluding phase is M (mitotic phase). The first three phases are together recognised as

Fractional Order Mathematical Model for the Cell Cycle of a Tumour Cell   133

"interphase". Let us write the number of interphase cells and mitotic cells at a time t as $\mathcal{V}(t)$ and $\mathcal{U}(t)$, respectively. The mathematical model for change in the number of cells is described by:

$$D_t \mathcal{Q}(t) = \mu_{\mathcal{Q}} \mathcal{V}(t) - (b_{\mathcal{Q}} + \mu_{G_0} \mathcal{Q}(t)), \tag{5.9}$$

$$D_t \mathcal{V}(t) = -(\mu_0 + \mu_{\mathcal{Q}})\mathcal{V}(t), \tag{5.10}$$

$$D_t \mathcal{U}(t) = 2b_1 \mathcal{U}(t) + b_{\mathcal{Q}} \mathcal{Q}(t) - \mu_1 \mathcal{U}(t). \tag{5.11}$$

here,
$\mu_{\mathcal{Q}}$ is the rate of transition from G_1 to G_0,
$b_{\mathcal{Q}}$ indicates rate of transition from G_0 to G_1,
μ_{G_0} denotes the death rate of G_0 cells,
b_1 stands for the division rate of M cells,
μ_1 indicates death rate M cells,
μ_0 denotes the death rate G_1 cells.

For introducing the memory effect previously mentioned in the mathematical formulation, we will proceed with replacement of the time derivative in equations (5.9) through (5.11), by the C-F differential operator of arbitrary order.

The fresh fractional model is:

$$D_t^\eta \mathcal{Q}(t) = \mu_{\mathcal{Q}} \mathcal{V}(t) - (b_{\mathcal{Q}} + \mu_{G_0} \mathcal{Q}(t)), \tag{5.12}$$

$$D_t^\eta \mathcal{V}(t) = -(\mu_0 + \mu_{\mathcal{Q}})\mathcal{V}(t), \tag{5.13}$$

$$D_t^\eta \mathcal{U}(t) = 2b_1 \mathcal{U}(t) + b_{\mathcal{Q}} \mathcal{Q}(t) - \mu_1 \mathcal{U}(t). \tag{5.14}$$

along the starting conditions: $\mathcal{Q}(0) = 2 \times 10^5$, $\mathcal{V}(0) = 1 \times 10^5$ and $\mathcal{U}(0) = 4 \times 10^5$.

5.4 Solution of the Model of the Cell Cycle for the Tumour Cell: Existence and Uniqueness

Let $\mathcal{H} = \mathcal{C}(I) \times \mathcal{C}(I) \times \mathcal{C}(I)$ be a Banach space and for $(\mathcal{Q}, \mathcal{V}, \mathcal{U}) \in \mathcal{H}$, $\|(\mathcal{Q}, \mathcal{V}, \mathcal{U})\| = \|\mathcal{Q}\| + \|\mathcal{V}\| + \|\mathcal{U}\|$, is the norm defined over it. Here, $\mathcal{Q}, \mathcal{V}, \mathcal{U} \in \mathcal{C}(I)$. $\mathcal{C}(I)$ is the space of real valued and continuous functions given on the interval I, and norm of $\mathcal{Y} \in \mathcal{C}(I)$ is $\|\mathcal{Y}\| = \sup\{|\mathcal{Y}(t)| : t \in I\}$.

Here, the fixed-point theorem is needed to show the existence of the solution for the system of equations (5.12) through (5.14). Employing the integral operator given in equation (5.6) of fractional order (studied by Losada and Nieto) on equations (5.12) through (5.14).

$$Q(t) - Q(0) = I_t^\eta \{\mu_Q V(t) - (b_Q + \mu_{G_0} Q(t))\}, \tag{5.15}$$

$$V(t) - V(0) = I_t^\eta \{-(\mu_0 + \mu_Q) V(t)\}, \tag{5.16}$$

$$U(t) - U(0) = I_t^\eta \{2b_1 U(t) + b_Q Q(t) - \mu_1 U(t)\}. \tag{5.17}$$

On using definition of I_t^η given by equation (5.6) we have,

$$Q(t) - Q(0) = \frac{2(1-\eta)}{(2-\eta)M(\eta)} \{\mu_Q V(t) - (b_Q + \mu_{G_0} Q(t))\}$$

$$+ \frac{2\eta}{(2-\eta)M(\eta)} \int_0^t \{\mu_Q V(s) - (b_Q + \mu_{G_0} Q(s))\} ds, \tag{5.18}$$

$$V(t) - V(0) = \frac{2(1-\eta)}{(2-\eta)M(\eta)} \{-(\mu_0 + \mu_Q) V(t)\}$$

$$+ \frac{2\eta}{(2-\eta)M(\eta)} \int_0^t \{-(\mu_0 + \mu_Q) V(s)\} ds, \tag{5.19}$$

$$U(t) - U(0) = \frac{2(1-\eta)}{(2-\eta)M(\eta)} \{2b_1 U(t) + b_Q Q(t) - \mu_1 U(t)\}$$

$$+ \frac{2\eta}{(2-\eta)M(\eta)} \int_0^t \{2b_1 U(s) + b_Q Q(s) - \mu_1 U(s)\} ds. \tag{5.20}$$

Let us write,

$$K_1(t, Q) = \mu_Q V(t) - (b_Q + \mu_{G_0} Q(t)), \tag{5.21}$$

$$K_2(t, V) = -(\mu_0 + \mu_Q) V(t), \tag{5.22}$$

$$K_3(t, U) = 2b_1 U(t) + b_Q Q(t) - \mu_1 U(t). \tag{5.23}$$

Theorem 5.1

If the following inequalities hold, the kernels K_1, K_2, K_3 fulfil the conditions of contraction and the Lipschitz condition: $0 \leq (b_Q + \mu_{G_0}) < 1$, $0 \leq (\mu_0 + \mu_Q) < 1$ and $0 \leq (2b_1 - \mu_1) < 1$.

Proof

Let us start with K_1. Let Q and $(Q)_1$ represent two functions, then

$$\| K_1(t,Q) - K_1(t,(Q)_1) \| = \| b_Q + \mu_{G_0}(Q - (Q)_1) \|, \tag{5.24}$$

$$\| K_1(t,Q) - K_1(t,(Q)_1) \| = (b_Q + \mu_{G_0}) \| Q - (Q)_1 \|, \tag{5.25}$$

$$\| K_1(t,Q) - K_1(t,(Q)_1) \| = \gamma_1 \| Q - (Q)_1 \|. \tag{5.26}$$

Clearly, $\gamma_1 = b_Q + \mu_{G_0}$ and $\| Q(t) \|$ is a bounded function. Hence, the Lipschitz condition is satisfied for K_1, and it is a contraction if $0 \leq (b_Q + \mu_{G_0}) < 1$. Similarly, the remaining two cases satisfies the Lipschitz condition, that is,

$$\| K_2(t,V) - K_2(t,(V)_1) \| = \gamma_2 \| V - (V)_1 \|, \tag{5.27}$$

$$\| K_3(t,U) - K_3(t,(U)_1) \| = \gamma_3 \| U - U_1 \|. \tag{5.28}$$

and contraction if $0 \leq \gamma_2 < 1$ and $0 \leq \gamma_3 < 1$. □

On using the preceding kernels, the equation (5.18) becomes,

$$Q(t) = Q(0) + \frac{2(1-\eta)}{(2-\eta)M(\eta)} K_1(t,Q) + \frac{2\eta}{(2-\eta)M(\eta)} \int_0^t K_1(s,Q) ds, \tag{5.29}$$

$$V(t) = V(0) + \frac{2(1-\eta)}{(2-\eta)M(\eta)} K_2(s,V) + \frac{2\eta}{(2-\eta)M(\eta)} \int_0^t K_2(s,V) ds, \tag{5.30}$$

$$R^*(t) = U(0) + \frac{2(1-\eta)}{(2-\eta)M(\eta)} K_3(t,U) + \frac{2\eta}{(2-\eta)M(\eta)} \int_0^t K_3(s,U) ds. \tag{5.31}$$

Corresponding recursive formulas are given by

$$Q_{(n)} = \frac{2(1-\eta)}{(2-\eta)M(\eta)} K_1(t,Q_{(n-1)}) + \frac{2\eta}{(2-\eta)M(\eta)} \int_0^t K_1(s,Q_{(n-1)}) ds, \tag{5.32}$$

$$V_{(n)} = \frac{2(1-\eta)}{(2-\eta)M(\eta)} K_2(t,V_{(n-1)}) + \frac{2\eta}{(2-\eta)M(\eta)} \int_0^t K_2(s,V_{(n-1)}) ds, \tag{5.33}$$

$$U_{(n)} = \frac{2(1-\eta)}{(2-\eta)M(\eta)} K_3(t,U_{(n-1)}) + \frac{2\eta}{(2-\eta)M(\eta)} \int_0^t K_3(s,U_{(n-1)}) ds. \tag{5.34}$$

The initial conditions are: $Q_0 = Q(0)$, $V_0 = V(0)$, $U_0 = U(0)$.

The following expressions depict, respectively, the difference of the terms in equations (5.32) through (5.34), including their preceding term,

$$\phi_n(t) = Q_{(n)} - Q_{(n-1)}$$
$$= \frac{2(1-\eta)}{(2-\eta)M(\eta)}(K_1(t, Q_{(n-1)}) - K_1(t, Q_{(n-2)})) \quad (5.35)$$
$$+ \frac{2\eta}{(2-\eta)M(\eta)} \int_0^t (K_1(s, Q_{(n-1)}) - K_1(s, Q))ds,$$

$$\psi_n(t) = V_{(n)} - V_{(n-1)}$$
$$= \frac{2(1-\eta)}{(2-\eta)M(\eta)}(K_2(t, V_{(n-1)}) - K_2(t, V_{(n-2)})) \quad (5.36)$$
$$+ \frac{2\eta}{(2-\eta)M(\eta)} \int_0^t (K_2(s, V_{(n-1)}) - K_1(s, V))ds,$$

$$\xi_n(t) = U_{(n)} - U_{(n-1)}$$
$$= \frac{2(1-\eta)}{(2-\eta)M(\eta)}(K_3(t, U_{(n-1)}) - K_3(t, U_{(n-2)})) \quad (5.37)$$
$$+ \frac{2\eta}{(2-\eta)M(\eta)} \int_0^t (K_3(s, U_{(n-1)}) - K_3(s, U))ds.$$

Hence,

$$Q_{(n)}(t) = \sum_{p=0}^{n} \phi_p(t), \quad (5.38)$$

$$V_{(n)}(t) = \sum_{p=0}^{n} \psi_p(t), \quad (5.39)$$

$$U_{(n)}(t) = \sum_{p=0}^{n} \xi_p(t). \quad (5.40)$$

Further, we have

$$\|\phi_n(t)\| = \|Q_{(n)} - Q_{(n-1)}\| = \left\| \frac{2(1-\eta)}{(2-\eta)M(\eta)}(K_1(t, Q_{(n-1)}) - K_1(t, Q_{(n-2)})) \right.$$

$$\left. + \frac{2\eta}{(2-\eta)M(\eta)} \int_0^t (K_1(s, Q_{(n-1)}) - K_1(s, Q))ds \right\|. \quad (5.41)$$

Fractional Order Mathematical Model for the Cell Cycle of a Tumour Cell 137

On applying triangle inequality,

$$\| \mathcal{Q}_{(n)} - \mathcal{Q}_{(n-1)} \| \leq \frac{2(1-\eta)}{(2-\eta)M(\eta)} \|(\mathcal{K}_1(t,\mathcal{Q}_{(n-1)}) - \mathcal{K}_1(t,\mathcal{Q}_{(n-2)}))\|$$

$$+ \frac{2\eta}{(2-\eta)M(\eta)} \left\| \int_0^t (\mathcal{K}_1(s,\mathcal{Q}_{(n-1)}) - \mathcal{K}_1(s,\mathcal{Q}))ds \right\|. \tag{5.42}$$

For the reason that the kernels fulfils the Lipschitz condition, we obtain

$$\| \mathcal{Q}_{(n)} - \mathcal{Q}_{(n-1)} \| \leq \frac{2(1-\eta)}{(2-\eta)M(\eta)} \gamma_1 \| \mathcal{Q}_{(n-1)} - \mathcal{Q}_{(n-2)} \|$$

$$+ \frac{2\eta}{(2-\eta)M(\eta)} \gamma_1 \int_0^t \| \mathcal{Q}_{(n-1)} - \mathcal{Q}_{(n-2)} \| ds, \tag{5.43}$$

and hence

$$\| \phi_n(t) \| \leq \frac{2(1-\eta)}{(2-\eta)M(\eta)} \gamma_1 \| \phi_{n-1}(t) \| + \frac{2\eta}{(2-\eta)M(\eta)} \gamma_1 \int_0^t \| \phi_{n-1}(s) \| ds. \tag{5.44}$$

Similarly,

$$\| \psi_n(t) \| \leq \frac{2(1-\eta)}{(2-\eta)M(\eta)} \gamma_1 \| \psi_{n-1}(t) \| + \frac{2\eta}{(2-\eta)M(\eta)} \gamma_1 \int_0^t \| \psi_{n-1}(s) \| ds \tag{5.45}$$

$$\| \xi_n(t) \| \leq \frac{2(1-\eta)}{(2-\eta)M(\eta)} \gamma_1 \| \xi_{n-1}(t) \| + \frac{2\eta}{(2-\eta)M(\eta)} \gamma_1 \int_0^t \| \xi_{n-1}(s) \| ds. \tag{5.46}$$

Theorem 5.2

In equations (5.12) through (5.14), modelling the cell cycle for a tumoural cell has exact coupled-solution subject to the condition that a t_0 can be found so that

$$\frac{2(1-\eta)}{(2-\eta)M(\eta)} \gamma_1 + \frac{2\eta}{(2-\eta)M(\eta)} \gamma_1 t_0 < 1. \tag{5.47}$$

Proof

Since the functions $\mathcal{Q}, \mathcal{V}, \mathcal{U}$ are bounded and the Lipschitz condition is fulfilled by the kernels from equations (5.44) and (5.45) and the recursive method, the subsequent relation can be determined as,

$$\| \phi_n(t) \| \leq \| \mathcal{Q}_{(n)}(0) \| \left[\left(\frac{2(1-\eta)}{(2-\eta)M(\eta)} \gamma_1 \right) + \left(\frac{2\eta}{(2-\eta)M(\eta)} \gamma_1 \right) \right]^n, \tag{5.48}$$

$$\|\psi_n(t)\| \leq \|\mathcal{V}_{(n)}(0)\| \left[\left(\frac{2(1-\eta)}{(2-\eta)M(\eta)}\gamma_2\right) + \left(\frac{2\eta}{(2-\eta)M(\eta)}\gamma_2\right)\right]^n, \quad (5.49)$$

$$\|\xi_n(t)\| \leq \|\mathcal{U}_{(n)}(0)\| \left[\left(\frac{2(1-\eta)}{(2-\eta)M(\eta)}\gamma_3\right) + \left(\frac{2\eta}{(2-\eta)M(\eta)}\gamma_1\right)\right]^n. \quad (5.50)$$

Therefore, the aforementioned functions will exist and are continuous.

Now, to demonstrate that these functions form a solution of equations (5.12) through (5.14), consider

$$\mathcal{Q}(t) - \mathcal{Q}(0) = \mathcal{Q}_{(n)} - A_n(t), \quad (5.51)$$

$$\mathcal{V}(t) - \mathcal{V}(0) = \mathcal{V}_{(n)} - B_n(t), \quad (5.52)$$

$$\mathcal{U}(t) - \mathcal{U}(0) = \mathcal{U}_{(n)} - C_n(t). \quad (5.53)$$

We obtain,

$$\|A_n(t)\| = \left\| \frac{2(1-\eta)}{(2-\eta)M(\eta)}(\mathcal{K}(t,\mathcal{Q}) - \mathcal{K}(t,\mathcal{Q}_{(n-1)})) \right.$$

$$\left. + \frac{2\eta}{(2-\eta)M(\eta)} \int_0^t (\mathcal{K}_1(s,\mathcal{Q}) - \mathcal{K}_1(s,\mathcal{Q}_{(n-1)}))ds \right\|,$$

$$\leq \frac{2(1-\eta)}{(2-\eta)M(\eta)} \|(\mathcal{K}(t,\mathcal{Q}) - \mathcal{K}(t,\mathcal{Q}_{(n-1)}))\|$$

$$+ \frac{2\eta}{(2-\eta)M(\eta)} \int_0^t \|(\mathcal{K}_1(s,\mathcal{Q}) - \mathcal{K}_1(s,\mathcal{Q}_{(n-1)}))\| ds,$$

$$\leq \frac{2(1-\eta)}{(2-\eta)M(\eta)}\gamma_1 \|\mathcal{Q} - \mathcal{Q}_{(n-1)}\| + \frac{2\eta}{(2-\eta)M(\eta)}\gamma_1 \|\mathcal{Q} - \mathcal{Q}_{(n-1)}\| t. \quad (5.54)$$

On repeated use of the preceding process,

$$\|A_n(t)\| \leq \left(\frac{2(1-\eta)}{(2-\eta)M(\eta)} + \frac{2\eta t}{(2-\eta)M(\eta)}\right)^{n+1} (\gamma_1)^{n+1} \mathcal{Q}(0). \quad (5.55)$$

At t_0,

$$\|A_n(t)\| \leq \left(\frac{2(1-\eta)}{(2-\eta)M(\eta)} + \frac{2\eta t_0}{(2-\eta)M(\eta)}\right)^{n+1} (\gamma_1)^{n+1} \mathcal{Q}(0). \quad (5.56)$$

Taking limit n tending to infinity in equation (5.57), $\|A_n(t)\| \to 0$.
Correspondingly, $\|B_n(t)\| \to 0, \|C_n(t)\| \to 0$.
This proves that the solution exists.

Now, to show that the solution is unique, let $\{Q^1, V^1, U^1\}$ be the other solution of equations (5.12) through (5.14).

$$Q(t) - Q^1(t) = \frac{2(1-\eta)}{(2-\eta)M(\eta)}(\mathcal{K}(t,Q) - \mathcal{K}_1(t,Q^1))$$

$$+ \frac{2\eta}{(2-\eta)M(\eta)}\int_0^t (\mathcal{K}(s,Q) - \mathcal{K}_1(s,Q^1))ds, \qquad (5.57)$$

Taking the norm in equation (5.58)

$$\|Q(t) - Q^1(t)\| \le \frac{2(1-\eta)}{(2-\eta)M(\eta)}\|\mathcal{K}(t,Q) - \mathcal{K}_1(t,Q^1)\|$$

$$+ \frac{2\eta}{(2-\eta)M(\eta)}\int_0^t \|\mathcal{K}(s,Q) - \mathcal{K}_1(s,Q^1)\| ds. \qquad (5.58)$$

On adopting the Lipschitz condition to the kernels, we have

$$\|Q(t) - Q^1(t)\| \le \frac{2(1-\eta)}{(2-\eta)(a)}\gamma_1 \|Q(t) - Q^1(t)\|$$

$$+ \frac{2\eta}{(2-\eta)M(\eta)}\gamma_1 t \|Q(t) - Q^1(t)\|. \qquad (5.59)$$

It gives

$$\|Q(t) - Q^1(t)\|\left(1 - \frac{2(1-\eta)}{(2-\eta)M(\eta)}\gamma_1 - \frac{2\eta}{(2-\eta)M(\eta)}\gamma_1 t\right) \le 0. \qquad (5.60)$$

With the help of equation (5.7), it can be easily shown that

$$\left(1 - \frac{2(1-\eta)}{(2-\eta)M(\eta)}\gamma_1 - \frac{2\eta}{(2-\eta)M(\eta)}\gamma_1 t\right) > 0. \qquad (5.61)$$

and if the preceding condition is observed, then

$$\|Q(t) - Q^1(t)\|\left(1 - \frac{2(1-\eta)}{(2-\eta)M(\eta)}\gamma_1 - \frac{2\eta}{(2-\eta)M(\eta)}\gamma_1 t\right) \le 0. \qquad (5.62)$$

$$\Rightarrow \|Q(t) - Q^1(t)\| = 0,$$

$$\Rightarrow Q(t) = Q^1(t).$$

Similarly, $V(t) = V^1(t)$, $U(t) = U^1(t)$.

This verifies uniqueness of the solution. □

5.5 Result and Discussion

In this segment, for the solution of the system in equations (5.12) through (5.14), the numerical simulation has been done. The solutions are obtained by the use of the iterative perturbation method, and it is advantageous to use this method because it provides an analytical approximation. Unlike the Adomain decomposition method, the perturbation iterative method is independent of Adomain polynomials. In this technique there is no requirement of the Lagrange multiplier, calculating integrals or correctional functionals; contrary to the homotopy perturbation method, this method does not require solution of the functional in each iteration. The solution of the model is obtained as follows:

For Q, we have

$$D_t^\eta Q(t) = \mu_Q V(t) - (b_Q + \mu_{G_0}) Q(t), \tag{5.63}$$

$$F(Q^n, Q, \dot{Q}, \epsilon) = \epsilon D_t^\eta Q(t) - \epsilon \mu_Q V(t) + \epsilon (b_Q + \mu_{G_0} Q(t)) + \dot{Q} - \epsilon \dot{Q}. \tag{5.64}$$

Now, assume an approximate solution of the system as,

$$Q_{(n+1)} = Q_{(n)} + \epsilon (Q_c)_n, \tag{5.65}$$

where subscript n represents the nth iteration over this approximate solution, and Q_c represents the correction term.

On approximating the system within a Taylor series expansion in the neighbourhood of $\epsilon = 0$ as

$$\dot{Q}_n + \epsilon(\dot{Q}_c)_n + \epsilon(D_t^\eta Q_n(t) - \mu_{Q_n} V_n(t) + (b_{Q_n} + \mu_{G_0}) Q_n(t)) - \dot{Q}_n = 0. \tag{5.66}$$

For the first approximation that corresponds to $n = 0$ and with the use of initial condition $Q(0) = Q_0 = 2 \times 10^5$, we obtain the following differential equation in Q_c

$$(\dot{Q}_c)_0 = \mu_{Q_0} V_0 - (b_{Q_0} + \mu_{G_0}) Q_0. \tag{5.67}$$

On substituting $(Q_c)_0 = 32002\,t$ in the recurrence relation in equation (5.65), we have

$$Q_1 = 2\times 10^5 - 32002\,t. \quad (5.68)$$

For $\mathcal{V}$

$$D_t^\eta \mathcal{V}(t) = -(\mu_0 + \mu_Q)\mathcal{V}(t), \quad (5.69)$$

$$F(\mathcal{V}^\eta, \mathcal{V}, \mathcal{V}, \varepsilon) = \varepsilon D_t^\eta \mathcal{V}(t) + \varepsilon(\mu_0 + \mu_Q)\mathcal{V}(t) + \dot{\mathcal{V}} - \varepsilon \dot{\mathcal{V}}. \quad (5.70)$$

Now, assume an approximate solution of the system as

$$\mathcal{V}_{(n+1)} = \mathcal{V}_{(n)} + \varepsilon(\mathcal{V}_c)_n, \quad (5.71)$$

where subscript n represents the nth iteration over this approximate solution, and $\mathcal{V}_c$ represents the correction term.

On approximating the system within a Taylor series expansion in the neighbourhood of $\varepsilon = 0$ as

$$\dot{\mathcal{V}}_n + \varepsilon((\dot{\mathcal{V}}_c)_n + \varepsilon(D_t^\eta \mathcal{V}_n) + (\mu_0 + \mu_Q)\mathcal{V}_n) - \dot{\mathcal{V}}_n = 0. \quad (5.72)$$

For the first approximation that corresponds to $n=0$ and with the use of initial condition $\mathcal{V}(0) = \mathcal{V}_0 = 4\times 10^5$, we obtain the following differential equation in $\mathcal{V}_c$

$$(\dot{\mathcal{V}}_c)_0 = (\mu_0 + \mu_{Q_0})\mathcal{V}_0. \quad (5.73)$$

On substituting $(\mathcal{V}_c)_0 = 52\times 10^3\,t$ in the recurrence relation in equation (5.66), we have

$$\mathcal{V}_1 = 4\times 10^5 + 52\times 10^3\,t. \quad (5.74)$$

For $\mathcal{U}$

$$D_t^\eta \mathcal{U}(t) = 2b_1 \mathcal{U}(t) + b_Q Q(t) - \mu_1 \mathcal{U}(t), \quad (5.75)$$

$$F(\mathcal{U}^\eta, \mathcal{U}, \mathcal{U}, \epsilon) = \epsilon D_t^\eta \mathcal{U} - 2b_1 \epsilon \mathcal{U} - b_Q \epsilon Q + \epsilon \mu_1 \mathcal{U} + \dot{\mathcal{U}} - \epsilon \dot{\mathcal{U}}. \quad (5.76)$$

Now, assume an approximate solution of the system as

$$\mathcal{U}_{(n+1)} = \mathcal{U}_{(n)} + \epsilon(\mathcal{U}_c)_n, \quad (5.77)$$

where subscript n represents the nth iteration over this approximate solution, and $\mathcal{U}_c$ represents the correction term.

On approximating the system within a Taylor series expansion in the neighbourhood of $\epsilon = 0$ as

$$\dot{\mathcal{U}}_n + \epsilon(\mathcal{U}_c)_n + \epsilon D_t^\eta \mathcal{U}_n - 2b_1\epsilon\mathcal{U}_n - b_{\mathcal{Q}_n}\epsilon\mathcal{Q}_n + \epsilon\mu_1\mathcal{U}_n - \dot{\mathcal{U}}_n = 0. \tag{5.78}$$

For the first approximation that corresponds to $n = 0$ and with the use of initial condition $\mathcal{U}(0) = \mathcal{U}_0 = 1 \times 10^5$, we obtain the following differential equation in $\mathcal{U}_c$

$$(\mathcal{U}_c)_0 = 2b_1\mathcal{U}_0 + b_{\mathcal{Q}_0}\mathcal{Q}_0 - \mu_1\mathcal{U}_0. \tag{5.79}$$

On substituting $(\mathcal{U}_c)_0 = 212 \times 10^3\, t$ in the recurrence relation in equation (5.77), we have

$$\mathcal{U}_1 = 10^5 + 212 \times 10^3\, t. \tag{5.80}$$

Similarly, the second set of approximations is as follows:

$$\mathcal{Q}_2 = 2 \times 10^5 - 32002 t + \left(\frac{32002}{a} - 32002\right) t$$
$$+ (520 + 20001 \times 0.18001) t^2 + 32002 \frac{(1-a)}{a^2} e^{-\frac{at}{(1-a)}}, \tag{5.81}$$

$$\mathcal{V}_2 = 4 \times 10^5 - \frac{52 \times 10^3}{a} t - 2730 t^2, \tag{5.82}$$

$$\mathcal{U}_2 = 10^5 + 452 \times 10^3 + \frac{212 \times 10^3 t}{a} + 208399.8 t^2 + 212 \times 10^3 \frac{(1-a)}{a^2} e^{-\frac{at}{1-a}}. \tag{5.83}$$

The graphs are made for the preceding solution with following values of the parameters [39]: $\mu_{G_0} = 0.1 \times 10^{-4}$ day^{-1}, $\mu_{\mathcal{Q}} = 0.02$ day^{-1}, $b_{\mathcal{Q}} = 0.2$ day^{-1}, $b_1 = 1$ day^{-1}, $\mu_0 = 0.11$ day^{-1} and $\mu_1 = 0.28$ day^{-1}. The initial conditions are: $\mathcal{Q}(0) = 2 \times 10^5$, $\mathcal{V}(0) = 1 \times 10^4$ and $\mathcal{U}(t) = 4 \times 10^5$.

Figures 5.1 through 5.3 showing the investigation for $\mathcal{Q}$, $\mathcal{V}$, and $\mathcal{U}$ for different values of "a". In Figure 5.1, we can observe that the quiescent cells, which are neither dividing nor preparing to divide, are decreasing with time and also increasing in value of the number of quiescent cells decreasing. Figure 5.2 is showing the number of interphase cells

Fractional Order Mathematical Model for the Cell Cycle of a Tumour Cell

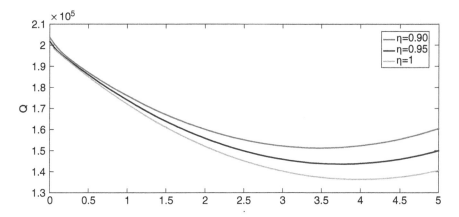

FIGURE 5.1
Graph between Q and t for numerous values of η.

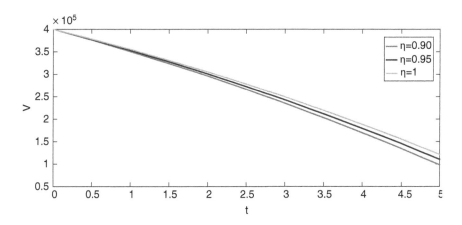

FIGURE 5.2
Graph between V and t for different values of η.

versus time, and as the process of the cell cycle gets started whenever the time (t) is increasing, then the number of interphase cells have to be decreased. Figure 5.3 reveals, as time increases, the number of mitotic cells increasing.

Suppose for treatment the system is employed to the immunotherapy gap between one mitosis and subsequently is sufficiently wide, and the division rate is small, then the tumour dies out independently of the delivered dose.

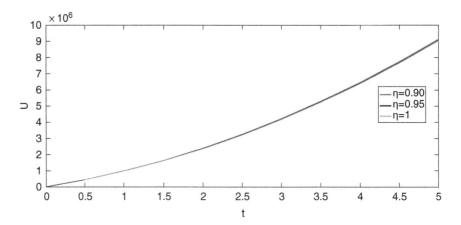

FIGURE 5.3
Graph between $\mathcal{U}$ and t for various values of η.

Nonetheless, cancer is by reason of undisciplined growth of cells, and accordingly, a large division is credible. In further effects of the constant immunotherapy the interphase time is large enough that slow down the growth and division of cell.

5.6 Conclusion

Here, the Caputo-Fabrizio fractional differential operator is applied, and the fractional model of the cell cycle for the tumour cell is investigated. The fixed-point theorem has been used to demonstrate the existence and uniqueness of the solution. The model solution is acquired by employing the perturbation iterative method. Some numerical simulation also has been completed, which represents the influence of the fractional order. Thus, giving the solution for the mathematical model of tumour growth is useful to analyse the tumour growth model.

References

1. Dyson, J., Villella-Bressan, R., and Webb, G. (2002). Asynchronous exponential growth in an age structured population of proliferating and quiescent cells. *Mathematical Biosciences*, 177:73–83.

2. Dyson, J., Villella-Bressan, R., and Webb, G. (2007). A spatial model of tumor growth with cell age, cell size, and mutation of cell phenotypes. *Mathematical Modelling of Natural Phenomena*, 2(3):69–100.
3. Liu, W., Hillen, T., and Freedman, H. (2007). A mathematical model for m-phase specific chemotherapy including the G_0-phase and immunoresponse. *Mathematical Biosciences and Engineering*, 4(2):239.
4. Villasana, M. and Radunskaya, A. (2003). A delay differential equation model for tumor growth. *Journal of Mathematical Biology*, 47(3):270–294.
5. Mackey, M. C. (2001). Cell kinetic status of haematopoietic stem cells. *Cell Proliferation*, 34(2):71–83.
6. Webb, G. F. (1992). A cell population model of periodic chemotherapy treatment. *Biomedical Modeling and Simulation*, pp. 83–92.
7. Kheifetz, Y., Kogan, Y., and Agur, Z. (2006). Long-range predictability in models of cell populations subjected to phase-specific drugs: Growth-rate approximation using properties of positive compact operators. *Mathematical Models and Methods in Applied Sciences*, 16(supp01):1155–1172.
8. Tubiana, M. and Malesia, E. (1976). Comparison of cell proliferation kinetics in human and experimental tumours: Response to irradiation. *Cancer Treatment Reports*, 60(12):1887–1895.
9. Agarwal, R., Kritika and Purohit, S. D. (2019a). A mathematical fractional model with nonsingular kernel for thrombin receptor activation in calcium signalling. *Mathematical Methods in the Applied Sciences*, 42(18):7160–7171.
10. Agarwal, R., Yadav, P. M., Agarwal, R., and Goyal, R. (2019b). Analytic solution of fractional advection dispersion equation with decay for contaminant transport in porous media. *Matematicki Vesnik*, 71(1):5–15.
11. Caputo, M. and Mainardi, F. (1971). A new dissipation model based on memory mechanism. *Pure and Applied Geophysics*, 91(1):134–147.
12. Drapaca, C. and Sivaloganathan, S. (2012). A fractional model of continuum mechanics. *Journal of Elasticity*, 107(2):105–123.
13. Ionescu, C., Lopes, A., Copot, D., Machado, J. T., and Bates, J. (2017). The role of fractional calculus in modeling biological phenomena: A review. *Communications in Nonlinear Science and Numerical Simulation*, 51:141–159.
14. Kumar, D., Singh, J., and Baleanu, D. (2016a). Numerical computation of a fractional model of differential-difference equation. *Journal of Computational and Nonlinear Dynamics*, 11(6):061004.
15. Kumar, S., Kumar, A., and Baleanu, D. (2016b). Two analytical methods for time-fractional nonlinear coupled Boussinesq–Burger's equations arise in propagation of shallow water waves. *Nonlinear Dynamics*, 85(2):699–715.
16. Lazopoulos, K. (2006). Non-local continuum mechanics and fractional calculus. *Mechanics Research Communications*, 33(6):753–757.
17. Luchko, Y. (2016). A new fractional calculus model for the two-dimensional anomalous diffusion and its analysis. *Mathematical Modelling of Natural Phenomena*, 11(3):1–17.
18. Nyamoradi, N., Zhou, Y., et al. (2015). Existence of solutions for a Kirchhoff type fractional differential equations via minimal principle and morse theory. *Topological Methods in Nonlinear Analysis*, 46(2):617–630.
19. Owolabi, K. M. and Atangana, A. (2017). Numerical approximation of nonlinear fractional parabolic differential equations with Caputo–Fabrizio derivative in Riemann–Liouville sense. *Chaos, Solitons & Fractals*, 99:171–179.

20. Podlubny, I. (1998). *Fractional Differential Equations: An Introduction to Fractional Derivatives, Fractional Differential equations, to Methods of Their Solution and some of Their Applications*, volume 198. Elsevier, Amsterdam, the Netherlands.
21. Sumelka, W., Blaszczyk, T., and Liebold, C. (2015). Fractional Euler–Bernoulli beams: Theory, numerical study and experimental validation. *European Journal of Mechanics-A/Solids*, 54:243–251.
22. Sumelka, W. and Voyiadjis, G. Z. (2017). A hyperelastic fractional damage material model with memory. *International Journal of Solids and Structures*, 124:151–160.
23. Wang, J., Fěckan, M., and Zhou, Y. (2017). Center stable manifold for planar fractional damped equations. *Applied Mathematics and Computation*, 296:257–269.
24. Zhou, Y. and Peng, L. (2017a). On the time fractional Navier–Stokes equations. *Applied Mathematics and Computation*, 73:874–891.
25. Zhou, Y. and Peng, L. (2017b). Weak solutions of the time-fractional Navier–Stokes equations and optimal control. *Computers & Mathematics with Applications*, 73(6):1016–1027.
26. Bagley, R. L. and Torvik, P. (1983). A theoretical basis for the application of fractional calculus to viscoelasticity. *Journal of Rheology*, 27(3):201–210.
27. Agarwal, R., Jain, S., and Agarwal, R. P. (2018). Mathematical modeling and analysis of dynamics of cytosolic calcium ion in astrocytes using fractional calculus. *Journal of Fractional Calculus and Applications*, 9(2):1–12.
28. Arshad, S., Baleanu, D., Bu, W., and Tang, Y. (2017). Effects of HIV infection on CD4+ T-cell population based on a fractional-order model. *Advances in Difference Equations*, 2017(1):92.
29. Losada, J. and Nieto, J. J. (2015). Properties of a new fractional derivative without singular kernel. *Progress Fractional Differentiation Applications*, 1(2):87–92.
30. Atangana, A. and Alkahtani, B. S. T. (2015). Analysis of the Keller–Segel model with a fractional derivative without singular kernel. *Entropy*, 17(6):4439–4453.
31. Caputo, M. and Fabrizio, M. (2016). Applications of new time and spatial fractional derivatives with exponential kernels. *Progress Fractional Differentiation and Applications*, 2(2):1–11.
32. Gao, F. and Yang, X.-J. (2016). Fractional Maxwell fluid with fractional derivative without singular kernel. *Thermal Science*, 20(3):871–877.
33. Gómez-Aguilar, J., Córdova-Fraga, T., Escalante-Martnez, J., Calderón-Ramón, C., and Escobar-Jiménez, R. (2016). Electrical circuits described by a fractional derivative with regular kernel. *Revista mexicana de fsica*, 62(2):144–154.
34. Hristov, J. (2016). Transient heat diffusion with a non-singular fading memory: From the Cattaneo constitutive equation with Jeffrey's kernel to the Caputo-Fabrizio time-fractional derivative. *Thermal Science*, 20(2):757–762.
35. Mirza, I. A. and Vieru, D. (2017). Fundamental solutions to advection–diffusion equation with time-fractional Caputo–Fabrizio derivative. *Computers & Mathematics with Applications*, 73(1):1–10.
36. Singh, J., Kumar, D., and Baleanu, D. (2018a). On the analysis of fractional diabetes model with exponential law. *Advances in Difference Equations*, 2018(1):231.

37. Singh, J., Kumar, D., Hammouch, Z., and Atangana, A. (2018b). A fractional epidemiological model for computer viruses pertaining to a new fractional derivative. *Applied Mathematics and Computation*, 316:504–515.
38. Caputo, M. and Fabrizio, M. (2015). A new definition of fractional derivative without singular kernel. *Progress in Fractional Differentiation and Applications*, 1(2):1–13.
39. Ravindran, N., Sheriff, M. M., and Krishnapriya, P. (2014). Homotopy perturbation method for solving cell cycle of tumoural cells. *British Journal of Mathematics and Computer Science*, 4(23):3271–3285.

6

Fractional Order Model of Transmission Dynamics of HIV/AIDS with Effect of Weak CD4+ T Cells

Ved Prakash Dubey, Rajnesh Kumar and Devendra Kumar

CONTENTS

6.1 Introduction 149
6.2 Mathematical Structure of the Fractional Order Model of Transmission Dynamics of HIV/AIDS with the Effect of Weak CD4+ T Cells 153
6.3 Preliminaries and Notations 155
6.4 Adam-Bashforth-Moulton Method for HIV Model 156
6.5 Numerical Results and Discussion 159
6.6 Concluding Remarks 161
References 162

6.1 Introduction

Acquired Immunodeficiency Syndrome (AIDS) is a vital problem in numerous emerging countries. It is an infectious disease of the human immune system induced by a retrovirus called the Human Immunodeficiency Virus (HIV). HIV spreads via certain body fluids that strike the body's immune system. Since CD4+ T cells are significant factors in the immune response, this is the main reason for the calamitous effect of HIV. HIV is actually such a type of retrovirus that attacks the CD4+ T lymphocytes that are the most abundant white blood cells of the immune system [1,2]. Despite the fact that HIV also contaminates other cells, it brings about the widespread annihilation of CD4+ T cells and thus lessens the resistivity of the immune system [3–5]. These special cells enhance the immune system to resist infections. CD4+ T cells are often called T cells, which are wiped out by this virus. Over a period of time, HIV breaks down many other cells, and the human body declines to resist these types of contaminations and disease.

At present, HIV/AIDS is a widespread problem around the entire world. At first, it was recognised in the world scenario by the US Centers for Disease

Control and Prevention in 1981. AIDS spreads out mainly through infected blood units or syringes, transfusion of infected blood, unprotected sexual relations, and from mother to her child during pregnancy. The vital transmission of HIV takes place through bodily fluids containing HIV, for instance, pre-seminal fluid, semen, blood, breast milk, and vaginal fluid [6]. Currently, HIV/AIDS is not curable and can be prevented only by avoiding contact with the virus.

A great interest has been shown by applied mathematicians to study the dynamics of HIV/AIDS. This mathematical research provides the appropriate treatment for infected humans to biologists. The dispersion and control of HIV/AIDS is modelled mathematically by way of short- and long-term forecasting of HIV and AIDS-related incidences. A mathematical model is an important tool to understand the transmission dynamics of the HIV/AIDS disease. In 1988, Anderson [7] propounded a model for HIV transmission, which divides the population class into many progressive stages of infection. Various mathematical models of non-linear nature have been established to depict the HIV dynamics, disease progression, and anti-retroviral response. In 1989, Perelson [8] established a model for HIV infection of the human immune system. The Perelson's model consists of three variables: the population size of infected cells, uninfected cells, and free virus particles. Perelson et al. [9] re-established a model provided by Perelson [8] by involving four variables: uninfected cells, latently infected cells, actively infected cells, and free virus particles. This model is defined by a set of four ordinary differential equations (ODEs). Culshaw and Ruan [10] minimised the model established by Perelson et al. [9] to a system of three ODEs with the assumption that all the infected cells can produce the virus. A mathematical model for infectious disease was also presented by Hethcote [11].

Mathematical models [12,13] have been utilised broadly to explore the dynamics of the epidemiology of HIV/AIDS. May and Anderson [7,14,15] have introduced the fundamental models related to HIV dynamics. Cai et al. [16] have used autonomous ODEs to develop an HIV/AIDS epidemic model with stage-structure analysis and treatment. A model of HIV/AIDS with the analysis of screening and treatment is described in [17]. There are also other biological models related to HIV in [18–21]. In 2009, a group of authors [22] have performed the modelling and analysis of the spread of AIDS epidemic with transference of HIV invectives. Mathematical modelling is also useful in the prediction of disease outbreaks. These models also evaluate the strategies related to prevention and drug therapy applied against the HIV-1 infection. Some researchers have introduced mathematical models [23–25] to investigate the transmission of HIV/AIDS due to sex workers.

During the past decade, a field of fractional calculus has attracted much attention from the research community to analyse the physical models describing the scientific phenomena. During the recent years, a bunch of mathematical models in the form of ODEs have been generated to describe the procedure of HIV infection of $CD4^+$ T cells because the spread of HIV infection is well-modelled with the set of differential equations. Zhou et al. [26] have modelled the HIV infection of CD4+ T cells with the cure rate in the form of differential equations. In 2011, various researchers have applied different approximate analytical techniques, such as the Laplace Adomian decomposition scheme [27], modified step differential transform method [28], homotopy analysis method [29,30] and variational iteration method [31] to derive the approximate solution of a fractional model of HIV infection of CD4+ T cells. In addition, the Bessel collocation scheme [32] and DTM [33] have been also utilised to investigate the numerical behaviour of a fractional HIV infection model. Javidi et al. [34] have examined the numerical behaviour of a fractional HIV epidemic model by using the Runge-Kutta method of the fourth order (RK4). Arafa et al. [35] have analysed the fractional order HIV model including the drug-therapy effect via the generalised Euler's scheme. An HIV model with various latent stages has been analysed by Huo et al. [36] regarding global stability. Homotopy decomposition scheme has been employed by Atangana et al. [37] for the computational analysis of the HIV infection model. In 2015, Khalid et al. [38] have applied the perturbation iteration method, a combination of perturbation extensions and Taylor series expansion, for the numerical solution of the HIV infection model. In 2014, Cai et al. [39] have analysed an extended HIV/AIDS epidemic model with treatment, and Huo et al. [40] have investigated its stability conditions.

In 2015, the non-standard finite difference scheme [41] has been employed to obtain the approximate analytical solution of the fractional model of HIV-1 infection of CD4+ T cells. In addition, the exponential Galerkin method [42] and the optimal variational iteration method [43] have also been employed for the fractional HIV model by some researchers. Mechee et al. [44] have applied the Lie symmetry approach to the mathematical model of HIV infection to evaluate the uninfected CD4+ T cells in the body. Recently, Pinto et al. [45] have propounded a fractional-order model for HIV dynamics with latency. The new thing of a latency fractional model of HIV is the inclusion of T-helper cells. Some authors have presented the comparison of numerical solutions of the fractional model of HIV infection via different approximating techniques. Al-Juaifri et al. [46] have provided the comparison analysis of homotopy perturbation method (HPM) and variational iteration method (VIM) regarding the mathematical model of HIV infection of CD4+ T cells. In this sequence, the semi-analytic schemes DTM and RK4 have been also used for comparison of solutions for the HIV/AIDS epidemic model by Olumuyiwa and Oluwaseun [47]. In 2019, Ali et al. [48] have examined the HIV model with latently infected cells through the Adomian decomposition method. Lichae et al. [49] introduced the fractional differential model of HIV-1 infection of CD4+ T cells with antiviral drug treatment effect and analysed numerically with the Laplace Adomian decomposition method. A Caputo-Fabrizio fractional order model for HIV infection has been investigated by Moore et al. [50] recently. In addition, Bulut et al. [51] presented an analytical study for a fractional HIV model describing infection of CD4+ T lymphocyte cells. Recently, a fractional order HIV model has been introduced and analysed to describe the effects of screening of unaware infectives by Babaei and Jafari [52].

In this sequence, some recent works can also be quoted. Samuel et al. [53] recently studied the effects of adrenal hormones on the immune system in respect of time evolution and spatial distribution cells. In addition, an investigation was also carried out by Samuel and Gill [54] to explore the effects of cortisol on immune response to HIV by deriving and analysing the diffusion-chemotaxis model, which also captures the role played by dendritic cells and other cells. There are multiple significant roles of dendritic cells at various stages of HIV infection identified by microbe sensing and communication with T cells. But their very close interaction with CD4+ T cells generates a place where infection of new targeted cells can occur. Regarding this fact, a time-fractional diffusion model introduced by Samuel and Gill [55] on dynamical effects of dendritic cells on HIV pathogenesis is very relevant for such a study.

The present work considers the application of the Adams-Bashforth-Moulton (ABM) method. The generalised ABM method, also termed as the fractional Adams method, is a proficient numerical scheme for handling fractional ODEs. Li and Tao [56] have provided fine literature on different aspects of the fractional ABM method. The ABM method actually allows us to explicitly compute the approximate solution at the present state from the solutions in previous states. Al-Sulami et al. [57] have investigated the fractional order Dengue epidemic model using generalised an ABM predictor-corrector scheme. In 2016–2017, the ABM scheme has been utilised successfully to numerically examine the fractional order susceptible-infectious-recovered model of Buruli ulcer disease [58] and bovine babesiosis disease [59]. Recently, the time-space fractional Keller-Segel chemotaxis system [60] and fractional Lengyel-Epstein chemical reaction model [61] have been analysed through fractional ABM scheme. In 2018, Sohail et al. [62] examined the stability properties for partial differential equations (PDEs) through a very new space spectral time-fractional ABM scheme. This scheme is built by combining the spectral scheme and the method of lines in which the system of fractional PDEs is transformed into ODEs.

In this paper, we study the mathematical model of the transmission dynamics of HIV/AIDS with the effect of weak CD4+ T cells recently established by Dutta and Gupta [63]. The ABM method has been used to solve and simulate the model. The central goal of the paper is to analyse the variations of uninfected cells, infected cells, and virus cells in respect of fractional order γ of time derivative and time t. The remaining part of the paper is structured as follows: In Section 6.2, we state the mathematical model of the transmission dynamics of HIV/AIDS. In Section 6.3, fundamental definitions and formulae regarding the fractional calculus are given. Section 6.4 presents the utilisation of the ABM method to obtain the numerical solutions of a fractional model. Section 6.5 is devoted to the numerical simulation and discussions for the model. In Section 6.6, we record the conclusions regarding the proposed study.

6.2 Mathematical Structure of the Fractional Order Model of Transmission Dynamics of HIV/AIDS with the Effect of Weak CD4+ T Cells

Dutta and Gupta [63] have proposed the following system of ODEs to model the transmission dynamics of HIV/AIDS with the effect of weak CD4+ T cells:

$$\frac{dT}{dt} = r - \alpha_1 VT - \alpha_2 VT - \beta_1 T + \alpha_3 I,$$

$$\frac{dI}{dt} = \alpha_2 VT - \alpha_3 I - \beta_2 I, \qquad (6.1)$$

$$\frac{dV}{dt} = A\beta_3 I + \alpha_1 VT - \beta_4 V,$$

with $T(0) = T_0 \geq 0$, $I(0) = I_0 \geq 0$, and $V(0) = V_0 \geq 0$, where:

$T(t)$, $I(t)$ and $V(t)$ represent the number of uninfected, infected, and virus CD4+ T cells, respectively
r denotes the inflow rate of CD4+ T cells
α_1 is the contact rate of CD4+ T cells and virus cells
$\alpha_2 (> \alpha_1)$ signifies the rate of infected cells
α_3 is the rate at which infected cells recover to uninfected cells
β_1 denotes the natural death rate of uninfected CD4+ T cells
$\beta_2 (> \beta_1)$ denotes the death rate of infected CD4+ T cells
β_3 indicates the lytic death rate for infected cells
β_4 indicates the death rate of virus cells
A is the average number of vital particles produced by an infected cell (Figure 6.1).

In this paper, the preceding System of equations (6.1) is considered with the fractional order γ as given below:

$$\frac{d^\gamma T}{dt^\gamma} = r - \alpha_1 VT - \alpha_2 VT - \beta_1 T + \alpha_3 I,$$

$$\frac{d^\gamma I}{dt^\gamma} = \alpha_2 VT - \alpha_3 I - \beta_2 I,$$

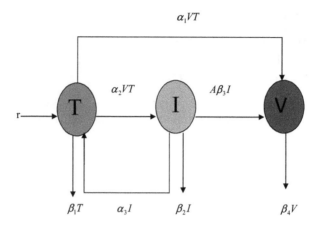

FIGURE 6.1
Transfer dynamics of the model (1).

$$\frac{d^\gamma V}{dt^\gamma} = A\beta_3 I + \alpha_1 VT - \beta_4 V,$$

with $T(0) = T_0 \geq 0$, $I(0) = I_0 \geq 0$, and $V(0) = V_0 \geq 0$.

Here $\frac{d^\gamma}{dt^\gamma}$ denotes the Caputo fractional derivative operator, and $0 < \gamma \leq 1$ specifies the fractional order of the time derivative (Table 6.1).

TABLE 6.1

Parameter Values in the HIV Model

Parameter Name	Notation	Value
Inflow rate of CD4+ T cells	r	10 mm^{-3}day^{-1}
The contact rate of CD4+ T cells and virus cells	α_1	0.000020 mm^{-3}day^{-1}
The rate of infected cells	α_2	0.000024 mm^{-3}day^{-1}
The rate at which infected cells recovers to uninfected cells	α_3	0.2 day^{-1}
The natural death rate of uninfected CD4+ T cells	β_1	0.01 day^{-1}
The death rate of infected CD4+ T cells	β_2	0.5 day^{-1}
The lytic death rate for infected cells	β_3	0.16 day^{-1}
The death rate of virus cells	β_4	3.4 day^{-1}
Average number of vital particles produced by an infected cell	A	1940

Source: Dutta, A. and Gupta, P. K., *Chin. J. Phys.*, 56, 1045–1056, 2018.

6.3 Preliminaries and Notations

This section presents the elemental definitions and formulae regarding the fractional derivatives and integrals in the field of fractional calculus.

Definition 6.1 [64]

A real valued function $\upsilon(\eta)$, $\eta > 0$ is said to be in the space $C_\sigma, \sigma \in \Re$ if $\exists$ a real number $\xi > \sigma$ such that $\upsilon(\eta) = \eta^\xi \upsilon_1(\eta)$ where $\upsilon_1 \in C[0, \infty)$ and exists in the space C_σ^m if $\upsilon^m \in C_\sigma, m \in \mathbb{N}$.

Definition 6.2 [64]

The Caputo fractional derivative of order ζ ($\zeta \geq 0$) is described by

$$D^\zeta \upsilon(\eta) = J^{\ell-\zeta} D^\ell \upsilon(\eta) = \frac{1}{\Gamma(\ell-\zeta)} \int_0^\eta (\eta-\xi)^{\ell-\zeta-1} \frac{d^\ell}{d\xi^\ell} \upsilon(\xi) d\xi,$$

$$\ell - 1 < \zeta \leq \ell, \ \eta > 0, \ \ell \in \mathbb{N},$$

where D^ℓ signifies the classical differential operator of order ℓ.

For the Caputo fractional derivative, we have

$$D^\zeta \tau^\gamma = \begin{cases} 0, & \gamma \leq \zeta - 1 \\ \dfrac{\Gamma(\gamma+1)}{\Gamma(\gamma-\zeta+1)} \tau^{\gamma-\zeta}, & \gamma > \zeta - 1. \end{cases}$$

Definition 6.3 [64]

The Caputo fractional derivative of order $\zeta > 0$ is given by

$$D_\eta^\zeta \upsilon(z,\eta) = \frac{\partial^\zeta \upsilon(z,\eta)}{\partial \eta^\zeta} = \begin{cases} \dfrac{1}{\Gamma(k-\zeta)} \int_0^\eta (\eta-\xi)^{k-\zeta-1} \dfrac{\partial^k \upsilon(z,\xi)}{\partial \xi^k} d\xi, & k-1 < \zeta < k \\ \dfrac{\partial^k \upsilon(z,\eta)}{\partial \eta^k}, & \zeta = k \in \mathbb{N}. \end{cases}$$

Definition 6.4 [65]

A fractional power series expansion about $\tau = \tau_0$ is given by

$$\sum_{l=0}^{\infty} d_l (\tau - \tau_0)^{l\beta} = d_0 + d_1 (\tau - \tau_0)^{\beta} + d_2 (\tau - \tau_0)^{2\beta} + \cdots,$$

$$0 \leq l-1 < \beta \leq l, \ \tau \geq \tau_0.$$

6.4 Adam-Bashforth-Moulton Method for HIV Model

The set of non-linear fractional order differential equations describing the transmission dynamics of HIV/AIDS with effect of weak CD4+ T cells is given as

$$\frac{d^{\gamma} T}{dt^{\gamma}} = r - \alpha_1 VT - \alpha_2 VT - \beta_1 T + \alpha_3 I$$

$$\frac{d^{\gamma} I}{dt^{\gamma}} = \alpha_2 VT - \alpha_3 I - \beta_2 I$$

$$\frac{d^{\gamma} V}{dt^{\gamma}} = A\beta_3 I + \alpha_1 VT - \beta_4 V \tag{6.2}$$

with $T(0) = T_0 \geq 0$, $I(0) = I_0 \geq 0$ and $V(0) = V_0 \geq 0$. (6.3)

This scheme is the generalisation of the ABM scheme. The approximate solutions of the set of non-linear fractional order differential equations (6.2) describing the transmission dynamics of the HIV model can be given by utilising this numerical algorithm as stated below.

The considered differential equation is

$$D_t^{\gamma} z(t) = v(t, z(t)), 0 \leq t \leq T,$$

$$z^{(l)}(0) = z_0^{(l)}, l = 0, 1, \ldots, [\gamma],$$

which is identical to the Volterra integral equation

$$z(t) = \sum_{l=0}^{[\gamma]-1} z_0^{(l)} \frac{t^l}{\Gamma(l)} + \frac{1}{\Gamma(\gamma)} \int_0^t (t-\tau)^{\gamma-1} v(s, z(\tau)) d\tau. \tag{6.4}$$

Setting $h = T/N$, $t_m = mh$, $m = 0, 1, 2, \ldots, N \in Z^+$.

Equation (6.4) can be discredited as

$$z_h(t_{m+1}) = \sum_{l=0}^{[\gamma]-1} z_0^{(l)} \frac{t_{m+1}^l}{\Gamma(l)} + \frac{h^\gamma}{\Gamma(\gamma+2)} \upsilon\left(t_{m+1}, z_h^q(t_{m+1})\right)$$

$$+ \frac{h^\gamma}{\Gamma(\gamma+2)} \sum_{j=0}^{m} \gamma_{j,m+1} \upsilon\left(t_h, z_h(t_j)\right),$$

$$\gamma_{j,m+1} = \begin{cases} m^{\gamma+1} - (m-\gamma)(m+1)^\gamma & , \text{if } j = 0, \\ (m-j+2)^{\gamma+1} + (m-j)^{\gamma+1} - 2(m-j+1)^{\gamma+1} & , \text{if } 0 \le j \le m, \\ 1 & , \text{if } j = m+1, \end{cases}$$

where the predicted value $z_h(t_{m+1})$ is calculated by

$$z_h^q(t_{m+1}) = \sum_{l=0}^{[\gamma]-1} z_0^{(l)} \frac{t_{m+1}^l}{\Gamma(l)} + \frac{1}{\Gamma(\gamma)} \sum_{j=0}^{m} b_{j,m+1} \upsilon\left(t_j, z_h(t_j)\right),$$

where $b_{j,m+1} = \frac{h^\gamma}{\gamma}((m+1-j)^\gamma - (m-j)^\gamma)$.

The error estimate is given by

$$\max_{j=0,1,\dots,N} \left| z(t_j) - z_h(t_j) \right| = 0(h^q) \text{ in which } q = \min(2, 1+\gamma).$$

By applying the preceding procedure, the system of equations (6.2) can be discredited as

$$T_{m+1} = T_0 + \frac{h^{\gamma_1}}{\Gamma(\gamma_1+2)} \left[r - (\alpha_1 + \alpha_2) V_{m+1}^q T_{m+1}^q - \beta_1 T_{m+1}^q + \alpha_3 I_{m+1}^q \right]$$

$$+ \frac{h^{\gamma_1}}{\Gamma(\gamma_1+2)} \sum_{j=0}^{m} \gamma_{1,j,m+1} \left[r - (\alpha_1 + \alpha_2) V_j T_j - \beta_1 T_j + \alpha_3 I_j \right],$$

$$I_{m+1} = I_0 + \frac{h^{\gamma_2}}{\Gamma(\gamma_2+2)} \left[\alpha_2 V_{m+1}^q T_{m+1}^q - (\alpha_3 + \beta_2) I_{m+1}^q \right]$$

$$+ \frac{h^{\gamma_2}}{\Gamma(\gamma_2+2)} \sum_{j=0}^{n} \gamma_{2,j,m+1} \left[\alpha_2 V_j T_j - (\alpha_3 + \beta_2) I_j \right],$$

$$V_{m+1} = V_0 + \frac{h^{\gamma_3}}{\Gamma(\gamma_3+2)}\left[A\beta_3 I_{m+1}^q + \alpha_1 V_{m+1}^q T_{m+1}^q - \beta_4 V_{m+1}^q\right]$$

$$+ \frac{h^{\gamma_3}}{\Gamma(\gamma_3+2)} \sum_{j=0}^{n} \gamma_{3,j,m+1}\left[A\beta_3 I_j + \alpha_1 V_j T_j - \beta_4 V_j\right],$$

in which

$$T_{m+1}^q = T_0 + \frac{1}{\Gamma(\gamma_1)} \sum_{j=0}^{n} B_{1,j,m+1}\left[r - (\alpha_1+\alpha_2)V_j T_j - \beta_1 T_j + \alpha_3 I_j\right],$$

$$I_{m+1}^q = I_0 + \frac{1}{\Gamma(\gamma_2)} \sum_{j=0}^{n} B_{2,j,m+1}\left[\alpha_2 V_j T_j - (\alpha_3+\beta_2)I_j\right],$$

$$V_{m+1}^q = V_0 + \frac{1}{\Gamma(\gamma_3)} \sum_{j=0}^{n} B_{3,j,m+1}\left[A\beta_3 I_j + \alpha_1 V_j T_j - \beta_4 V_j\right],$$

$$\gamma_{i,j,m+1} = \begin{cases} m^{\gamma_i+1} - (m-\gamma_i)(m+1)^{\gamma_i} & ,if\ j=0, \\ (m-j+2)^{\gamma_i+1} + (m-j)^{\gamma_i+1} - 2(m-j+1)^{\gamma_i+1} & ,if\ 0 \leq j \leq m, \\ 1 & ,if\ j = m+1, \end{cases}$$

$$B_{i,j,m+1} = \frac{h^{\gamma_i}}{\gamma_i}\left[(m+1-j)^{\gamma_i} - (m-j)^{\gamma_i}\right], 0 \leq j \leq m \text{ and } i = 1, 2, 3.$$

Using these schemes, we plot the graph of uninfected CD4+ T cells $T(t)$, HIV infected cells $I(t)$, and the virus cells $V(t)$ with respect to time t and fractional parameter γ.

The enticing characteristics of an ABM method are the computation of a local truncation error and the possibility of an inclusion of a correction term. These two attributes authenticate the accuracy of the calculation at each step in the computational process of the ABM method. But this numerical method has also some limitations in comparison to single-step methods like Euler, Taylor, and Runge-Kutta. These methods utilise only the data from one preceding point to compute the successive point, that is, there is a requirement of only the initial point, which will be used to compute the successive point, and in general the kth term is needed to compute $(k+1)$th term. But the ABM method is not a self-starting method. It requires several prior terms in the computational procedure to compute the successive term.

6.5 Numerical Results and Discussion

In this section, numerical results of the uninfected, infected, and virus CD4+ T cells for various values of fractional parameter $\gamma = 0.5, 0.75$ and for the standard value $\gamma = 1$ are calculated for different values of parameter r that denote the inflow rate of CD4+ T cells. The results are presented graphically through Figures 6.2–6.4. For $r = 10$, Figure 6.2a–c reports that it takes less time to meet uninfected and virus cells as the fractional parameter γ of time derivative decreases. It is also observed from Figure 6.2a–c that the number of infected and virus T cells decrease with increasing value of

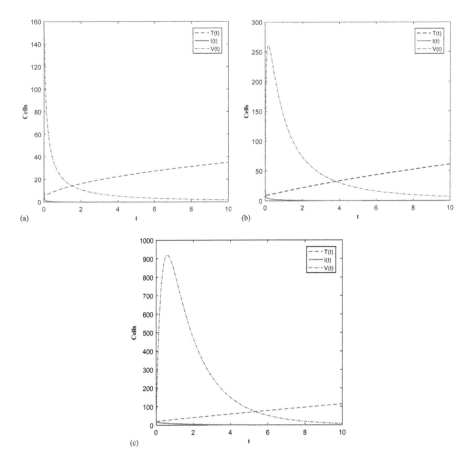

FIGURE 6.2
Plots of T, I, V vs. t with initial conditions $T(0) = 20$, $I(0) = 15$, $V(0) = 10$ and parameter values $r = 10$, $\alpha_1 = 0.000020$, $\alpha_2 = 0.000024$, $\alpha_3 = 0.2, \beta_1 = 0.01$, $\beta_2 = 0.5$, $\beta_3 = 0.16$, $\beta_4 = 3.4$, $A = 1940$: (a) $\gamma = 0.5$, (b) $\gamma = 0.75$, and (c) $\gamma = 1$.

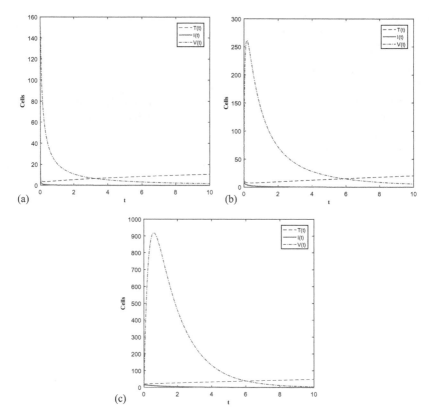

FIGURE 6.3
Plots of T, I, V vs. t with initial conditions $T(0) = 20$, $I(0) = 15$, $V(0) = 10$ and parameter values $r = 3$, $\alpha_1 = 0.000020$, $\alpha_2 = 0.000024$, $\alpha_3 = 0.2$, $\beta_1 = 0.01$, $\beta_2 = 0.5$, $\beta_3 = 0.16$, $\beta_4 = 3.4$, $A = 1940$: (a) $\gamma = 0.5$, (b) $\gamma = 0.75$, and (c) $\gamma = 1$.

time t and decreasing value of γ, while a number of uninfected cells continuously increase with increasing value of time t and relatively decrease with decreasing value of γ. Figure 6.2a–c also depicts that as the value of γ decreases, the lines representing uninfected and infected cells become closer to each other. It indicates that the rate of infection enhances as the fractional parameter γ decreases. The results from Figure 6.3a–c for $r = 3$ can be explained in a similar way.

Figure 6.4a–c shows that as the value of γ decreases, the lines representing uninfected and virus cells become much closer to each other. Figure 6.4a–c also depicts that as the value of γ decreases the lines representing uninfected and infected cells become closer to each other. The population of uninfected cells decreases with increasing value of time and decreasing value of γ.

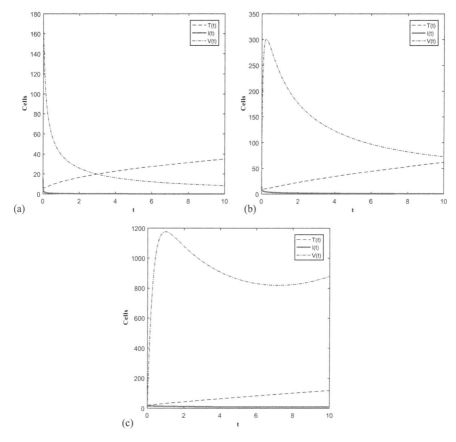

FIGURE 6.4
Plots of T, I, V vs. t with initial conditions $T(0) = 20$, $I(0) = 15$, $V(0) = 10$ and parameter values $r = 10$, $\alpha_1 = 0$, $\alpha_2 = 0.000024$, $\alpha_3 = 0.2$, $\beta_1 = 0.01$, $\beta_2 = 0$, $\beta_3 = 0.16$, $\beta_4 = 3.4$, $A = 1940$: (a) $\gamma = 0.5$, (b) $\gamma = 0.75$, and (c) $\gamma = 1$.

6.6 Concluding Remarks

In this study, the ABM method is applied to the fractional mathematical model of HIV/AIDS, including the effect of weak CD4+ T cells to derive its numerical solution. The numerical solutions have been utilised to analyse the variations of uninfected cells, infected cells, and virus cells with respect to time and different values of fractional order of the time derivative. The most consequential part of this study is the analysis of the time requirement for the contact of various categories of cells with variations of fractional order of the time derivative. These variations are exhibited through various graphs.

References

1. A.A.M. Arafa, S.Z. Rida, M. Khalil, Fractional modelling dynamics of HIV and CD4+ T-cells during primary infection, *Nonlinear Biomed. Phys.* 6(1) (2012) 1–7.
2. L. Wang, M.Y. Li, Mathematical analysis of the global dynamics of a model for HIV infection of CD4+ T cells, *Math. Biosci.* 200(1) (2006) 44–57.
3. L. Rong, M.A. Gilchrist, Z. Feng, A.S. Perelson, Modeling within host HIV-1 dynamics and the evolution of drug resistance: Tradeoffs between viral enzyme function and drug susceptibility, *J. Theor. Biol.* 247(4) (2007) 804–818.
4. L. Rong, Z. Feng, A.S. Perelson, Emergence of HIV-1 drug resistance during antiretroviral treatment, *Bull. Math. Biol.* 69(6) (2007) 2027–2060.
5. P.K. Srivastava, M. Banerjee, P. Chandra, Modelling the drug therapy for HIV infection, *J. Biol. Syst.* 17(2) (2009) 213–223.
6. Centers for Disease Control and Prevention (CDC). 2003. Atlanta, U.S. Department of Health and Human Services, https://www.cdc.gov/
7. R.M. Anderson, The role of mathematical models in the study of HIV transmission and the epidemiology of AIDS, *J. AIDS* 1 (1988) 241–256.
8. A.S. Perelson, Modelling the interaction of the immune system with HIV, in: C. Castillo-Chavez (Ed.), *Mathematical and Statistical Approaches to AIDS Epidemiology*, Springer, Berlin (1989) 350–370.
9. A.S. Perelson, D.E. Kirschner, R. De Boer, Dynamics of HIV infection of CD4+ T-cells, *Math. Biosci.* 114(1) (1993) 81–125.
10. R.V. Culshaw, S. Ruan, A delay-differential equation model of HIV infection of CD4+ T-cells, *Math. Biosci.* 165(1) (2000) 27–39.
11. H.W. Hethcote, The mathematics of infectious disease, *SIAM Rev.* 42(4) (2000) 599–653.
12. Y. Ding, H. Ye, A fractional-order differential equation model of HIV infection of CD4+ T-cells, *Math. Comput. Model.* 50(3–4) (2009) 386–392.
13. P.W. Nelson, A.S. Perelson, Mathematical analysis of delay differential equation models of HIV-1 infection, *Math. Biosci.* 179(1) (2002) 73–94.
14. R.M. Anderson, G.F. Medly, R.M. May, A.M. Johnson, A preliminary study of the transmission dynamics of the human immunodeficiency virus (HIV), the causative agent of AIDS, *IMA J. Math. Appl. Med. Biol.* 3(4) (1986) 229–263.
15. R.M. May, R.M. Anderson, Transmission dynamics of HIV infection, *Nature* 326 (1987) 137–142.
16. L. Cai, X. Li, M. Ghosh, B. Guo, Stability of an HIV/AIDS epidemic model with treatment, *J. Comput. Appl. Math.* 229(1) (2009) 313–323.
17. F. Nyabadza, Z. Mukandavire, S.D. Hove-Musekwa, Modelling the HIV/AIDS epidemic trends in South Africa: Insights from a simple mathematical model, *Nonlinear Anal. Real World Appl.* 12 (2011) 2091–2104.
18. R. Xu, Global stability of an HIV-1 infection model with saturation infection and intracellular delay, *J. Math. Anal. Appl.* 375(1) (2011) 75–81.
19. X. Song, A.U. Neumann, Global stability and periodic solution of the viral dynamics, *J. Math. Anal. Appl.* 329(1) (2007) 281–297.
20. X. Song, X. Zhou, X. Zhao, Properties of stability and Hopf bifurcation for a HIV infection model with time delay, *Appl. Math. Model.* 34(6) (2010) 1511–1523.
21. T. Zhang, M. Jia, H. Luo, Y. Zhou, N. Wang, Study on a HIV/AIDS model with application to Yunnan province, *Appl. Math. Model.* 35(9) (2011) 4379–4392.

22. R. Naresh, A. Tripathi, D. Sharma, Modelling and analysis of the spread of AIDS epidemic with immigration of HIV infectives, *Math. Comput. Model.* 49(5–6) (2009) 880–892.
23. R.M. Anderson, R.M. May, Population biology of infectious diseases: Part I, *Nature* 280 (1979) 361–367.
24. R.M. Anderson, Discussion: The Kermack-Mc Kendrick epidemic threshold theorem, *Bull. Math. Biol.* 53(1–2) (1991) 3–32.
25. N. Kaur, M. Ghosh, S.S. Bhatia, Mathematical analysis of the transmission dynamics of HIV/AIDS: Role of female sex workers, *Appl. Math. Inf. Sci.* 8(5) (2014) 2491–2501.
26. X. Zhou, X. Song, X. Shi, A differential equation model of HIV infection of CD4+ T cells with cure rate, *J. Math. Anal. Appl.* 342 (2008) 1342–1355.
27. M.Y. Ongun, The Laplace Adomian decomposition method for solving a model for HIV infection of CD4+ T cells, *Math. Comput. Model.* 53(5–6) (2011) 597–603.
28. A. Gökdoğan, A. Yildirim, M. Merdan, Solving a fractional order model of HIV infection of CD4+ T cells, *Math. Comput. Model.* 54(9–10) (2011) 2132–2138.
29. M. Ghoreishi, A.I.B. Md. Ismail, A.K. Alomari, Application of the homotopy analysis method for solving a model for HIV infection of CD4+ T cells, *Math. Comput. Model.* 54(11–12) (2011) 3007–3015.
30. P. Rani, D. Jain, V.P. Saxena, Approximate analytical solution with stability analysis of HIV/AIDS model, *Cogent Math.* 3(1) (2016) 1–14.
31. M. Merdan, A. Gökdoğan, A. Yildirim, On the numerical solution of the model for HIV infection of CD4+ T cells, *Comput. Math. Appl.* 62(1) (2011) 118–123.
32. S. Yüzbasi, A numerical approach to solve the model for HIV infection of CD4+ T cells, *Appl. Math. Model.* 36(12) (2012) 5876–5890.
33. V.K. Srivastava, M.K. Awasthi, S. Kumar, Numerical approximation for HIV infection of CD4+ T cells mathematical model, *AIN Shams Eng. J.* 5(2) (2014) 625–629.
34. M. Javidi, N. Nyamoradi, Numerical behaviour of a fractional order HIV/AIDS epidemic model, *World J. Model. Simul.* 9(2) (2013) 139–149.
35. A.A.M. Arafa, S.Z. Rida, M. Khalil, A fractional-order model of HIV infection with drug therapy effect, *J. Egypt. Math. Soc.* 22(3) (2014) 538–543.
36. H.-F. Huo, L.-X. Feng, Global stability for an HIV/AIDS epidemic model with different latent stages and treatment, *Appl. Math. Model.* 37(3) (2013) 1480–1489.
37. A. Atangana, E.F.D. Goufo, Computational analysis of the model describing HIV infection of CD4+ T-cells, *BioMed Res. Int.* (2014) 1–7.
38. M. Khalid, M. Sultana, F. Zaidi, F.S. Khan, A numerical solution of a model for HIV infection CD4+ T cells, *Int. J. Innov. Sci. Res.* 16(1) (2015) 79–85.
39. L. Cai, S. Guo, S. Wang, Analysis of an extended HIV/AIDS epidemic model with treatment, *Appl. Math. Comput.* 236 (2014) 621–627.
40. H.-F. Huo, R. Chen, X.-Y. Wang, Modelling and stability of HIV/AIDS epidemic model with treatment, *Appl. Math. Model.* 14(13–14) (2016) 6550–6559.
41. S. Zibaei, M. Namjoo, A nonstandard finite difference scheme for solving fractional order model of HIV-1 infection of CD4+ T-cells, *Iranian J. Math. Chem.* 6(2) (2015) 169–184.
42. S. Yüzbasi, M. Karaçayir, An exponential Galerkin method for solutions of HIV infection model of CD4+ T-Cells, *Comput. Bio. Chem.* 67 (2017) 205–212.
43. A. Nazir, N. Ahmed, U. Khan, S.T. Mohyud-Din, Analytical approach to study a mathematical model of CD4+ T cells, *Int. J. Biomath.* 11(4) (2018) 1–12.

44. M.S. Mechee, N. Haitham, Application of Lie symmetry for mathematical model of HIV infection of CD4+ T-cells, *Int. J. Appl. Eng. Res.* 13(7) (2018) 5069–5074.
45. C.M.A. Pinto, A.R.M. Carvalho, A latency fractional order model for HIV dynamics, *J. Comput. Appl. Math.* 312(C) (2017) 240–256.
46. G.A. Al-Juaifri, Z.I. Salman, M.S. Mechee, A comparison study of the mathematical model of HIV infection of CD4+ T-cells using homotopy perturbation and variational iteration methods, *Appl. Math. Sci.* 12(27) (2018) 1325–1340.
47. P.J. Olumuyiwa, A. Oluwaseun, Semi-analytic method for solving HIV/AIDS epidemic model, *Int. J. Mod. Biol. Med.* 9(1) (2018) 1–8.
48. N. Ali, S. Ahmad, S. Aziz, G. Zaman, The Adomian decomposition method for solving HIV infection model of latently infected cells, *Matrix Science Mathematic (MSMK)* 3(1) (2019) 05–08.
49. B.H. Lichae, J. Baizar, Z. Ayati, The fractional differential model of HIV-1 infection of CD4+ T cells with description of the effect of antiviral drug treatment, *Comput. Math. Methods Med.* 2019 (2019) 1–12.
50. E.J. Moore, S. Sirisubtawee, S. Koonprasert, A Caputo-Fabrizio fractional differential equation model for HIV/AIDS with treatment compartment, *Adv. Differ. Equ.* 200 (2019) 1–20.
51. H. Bulut, D. Kumar, J. Singh, R. Swroop, H.M. Baskonus, Analytic study for a fractional model of HIV infection of CD4+ T lymphocyte cells, *Math. Nat. Sci.* 2(1) (2018) 33–43.
52. A. Babaei, H. Jafari, A fractional order HIV/AIDS model based on the effect of screening of unaware infectives, *Math. Meth. Appl. Sci.* 42(7) (2019) 2334–2343.
53. S. Samuel, V. Gill, D. Kumar, Y. Singh, Numerical study of effects of adrenal hormones on lymphocytes, Mathematical Modelling, *Appl. Anal. Comput.* 272 (2019) 261–273.
54. S. Samuel, V. Gill, Diffusion-Chemotaxis model of effects of Cortisol on immune response to human immunodeficiency virus, *Nonlinear Eng.* 7(3) (2018) 207–227.
55. S. Samuel, V. Gill, Time-fractional diffusion model on dynamical effect of dendritic cells on HIV pathogenesis, *J. Comput. Methods Sci. Eng.* 18(1) (2018) 193–212.
56. C. Li, C. Tao, On the fractional Adams method, *Comput. Math. Appl.* 58(8) (2009) 1573–1588.
57. H. Al-Sulami, M. El-Shahed, J.J. Nieto, W. Shammakh, On fractional order Dengue epidemic model, *Math. Probl. Eng.* 2014 (2014) 1–6.
58. B. Ebenezer, G. Frimpong, A. Abubakari, Fractional order SIR model of Buruli ulcer disease, *Acad. J. Appl. Math. Sci.* 2(1) (2016) 1–10.
59. Z.A. Zafar, K. Rehan, M. Mushtaq, Fractional-order scheme for bovine babesiosis disease and tick populations, *Adv. Differ. Equ.* 86 (2017) 1–19.
60. M. Zayernouri, A. Matzavinos, Fractional Adams-Bashforth/Moulton methods: An application to the fractional Keller-Segel Chemotaxis system, *J. Comput. Phys.* 317 (2016) 1–14.
61. Z.A. Zafar, Fractional order Lengyel-Epstein chemical reaction model, *Comput. Appl. Math.* 38(3) (2019) 131.
62. A. Sohail, K. Maqbool, R. Ellahi, Stability analysis for fractional-order partial differential equations by means of space spectral time Adams-Bashforth Moulton method, *Numer. Meth Part Differ Equ.* 34(1) (2018) 19–29.

63. A. Dutta, P.K. Gupta, A mathematical model of transmission dynamics of HIV/AIDS with effect of weak CD4+ T cells, *Chin. J. Phys.* 56(3) (2018) 1045–1056.
64. I. Podlubny, *Fractional Differential Equations*, Academic Press, New York (1999).
65. H.M. Jaradat, S. Al-Shara, Q.J. Khan, M. Alquran, K. Al-Khaled, Analytical solution of time-fractional Drinfeld-Sokolov-Wilson system using residual power series method, *IAENG Int. J. Appl. Math.* 46(1) (2016) 64–70.

7

Fractional Dynamics of HIV-AIDS and Cryptosporidiosis with Lognormal Distribution

M. A. Khan and Abdon Atangana

CONTENTS

7.1 Introduction 167
7.2 Model Description 169
7.3 Basic Concepts of the CF and AB Fractional Derivative 171
7.4 HIV-AIDS and Cryptosporidiosis Fractional Model 174
    7.4.1 Existence and Uniqueness of HIV-AIDS
           and Cryptosporidiosis Fractional Model 174
7.5 HIV-AIDS and Cryptosporidiosis with the ABC-Derivative 184
    7.5.1 Existence of Solutions for the Co-infection Model................... 185
7.6 Numerical Approximations 187
7.7 Numerical Solution... 189
7.8 Model with Stochastic Approach: Application of Lognormal
    Distribution... 201
    7.8.1 Numerical Solution of Random Model for HIV-AIDS
           and Cryptosporidiosis 204
7.9 Conclusion 206
References... 206

7.1 Introduction

Due to the rate of death caused by the infectious diseases, researchers have devoted their attention to understand, analyse, and predict the behaviour of such diseases. More precisely, mathematicians rely on mathematical tools to construct mathematical models able to describe the spread of diseases in given targeted populations. In the last decade, one of the most fatal disease that have taken the lives of several people is perhaps HIV/AIDS. The disease is more present in less developed countries due to their lack of hygienic practices. Beside HIV/AIDS, we can also mention here that cryptospirodiosis is also one of the diseases in less developed countries that creates infection.

Common diseases in sub-Saharan Africa are malaria, schistosomiasis, and trypanosomiasis, and crytopspirodiosis is also one of the diseases in this area. It is caused by a microscopic parasite, Cryptosporidium, which lives inside the animal's or human's intestine and is passed in the stool of an infected person or animal, which causes the infection [1,2]. The disease as well as the parasite is known as 'crypto'. Globally, the species of cryptosporidium is responsible for the diarrhea infection in adults and children with a higher impact on chronic and life-threatening illness in immunocompromised patients, and especially with HIV/AIDS [3,4], in sub-Saharan Africa, Asia, and European continents.

The infection of HIV has become a major health and life-threating problem for the entire world, and particularly in the backward areas of central China, which is caused mainly due to the illegal blood collection and supply. The co-infection of Cryptospirodiosis and HIV covered large areas of the world [5]. The occupance of the co-infection of cryptospirodiosis and HIV co-infection has been documented in many studies, see for example [5–9] and the references therein. The cryptospirodiosis and HIV/AIDS and their co-infection leads to major illness in humans, so prevention and early diagnosis of patients in sensitive areas is recommended.

Fractional calculus has gained a lot of attention from researchers and scientists around the globe. It can best describe the real-world phenomenon rather than an ordinary derivative. It's application can bee seen in many areas of science and engineering [10–13]. The fractional models of real-life problems are considered important, and the reason is that they have the property of heredity and the effect of memory [14,15]. Further, the integer order models cannot explore well the dynamics between two different points. Due to such limitations of local differentiations, some new useful concepts on differentiations with non-local or fractional order are considered in the literature [15]. A few years ago in 2015, the Caputo-Fabrizio (CF)derivative was introduced with an exponential kernel by Caputo and Fabrizio [16]. The CF derivative has been applied successfully in a variety of real-life phenomena, see [17,30–36] and the references therein. One year later in 2016, a new fractional derivative, which generalises the Mittag-Leffler function as a non-local and non-singular kernel, was introduced by Atangana and his co-author Baleanu [37]. The concept of the Mittag-Leffler function as a non-local and non-singular kernel has been used numerously in literature, for example [38–40], and the references therein. Recently, the Ebola model has been analysed in the fractional Atangana-Baleanu (AB) derivative in [41]. Most recently, S. Ullah et al. considered the Hepatitis B virus (HBV) model with the AB derivative and successfully presented the dynamics of HBV [40].

The classical differential model is appropriate for the local dynamic system with no external forces. The model with the classical derivative cannot therefore replicate the complexity of the spread of the infectious disease model, as the model sometime has a crossover behaviour, and this cannot be handled by classical differential operators. The application of the CF and AB derivative to this model provides a novel model with crossover behaviour as was presented in the published works [17,40,42–45]. Some other related works and their applications to fractional calculus can be seen in [18–21], where efforts are shown to described the dynamics of a biological model through collocation methods, analytical solution of oxygen diffusion equations, the dynamics of cancer therapy, and the dynamics of Nagumo equations. Some more important works related to fractional calculus are studied in [22–29], where the authors studied the RLC circuit in fractional operators, physical interpretation of fractional calculus, application of fractional calculus to electromagnetic waves, solution of space/time distribution problems, solution of space/time fractional differential equations, analysis of the reaction diffusion system, new investigations of the CF operator and their applications to the multi-step homotopy analysis method and the synchronisation of fractional systems. Some more work related to the fractional derivativrs, see the work in [47–51].

In all of the above work, the fractional order derivative was applied to the real-world problems as a single operator, but no one applied both the new fractional derivatives at the same time for such infectious disease, that is the CF and the AB derivative. In the present work, we consider the co-infection model presented in [52] to investigate their fractional dynamics with the CF and AB derivative. Besides this, we also apply the lognormal and stochastic concept to the parameters and investigate their dynamics. We presented a detailed discussion on the proposed co-infection model, and then the application of the fractional calculus applied to real-life problems are presented above. To structure the rest of the work in this paper, we give the section-wise details: The co-infection model and their related discussion is presented in Section 7.2. The fractional calculus and its basic concept is given in Section 7.3. In Section 7.4, the CF operator is applied to the co-infection model and presents the existence and uniqueness of the model. The same model is formulated in the sense of the AB operator. Their existence and uniqueness is shown in Section 7.5. The numerical approximations for the AB derivative is presented in Section 7.6 while the numerical solution of the CF and AB fractional models is given in Section 7.7. In Section 7.8, we use the lognormal distribution concept and formulate as a novel model with numerical investigations. Finally, in Section 7.9, we summarise our work.

7.2 Model Description

Here, we considered the model of HIV/AIDS and crptospirodiosis presented in [52], to explore their dynamics in fractional order that is the CF and AB derivative and the novel concept of lognormal distributions. In order to investigate these new ideas to the co-infection model, first, we present the model and describe its variables and parameters in detail. The model that describes the co-dynamics of HIV/AIDS and cryptospirodiosis is given by:

$$\begin{cases} \dfrac{d}{dt}S = \Lambda + \sigma_1 R - \mu S - \beta_c^* S - \beta_H^* S \\[4pt] \dfrac{d}{dt}I = \beta_c^* S - (\alpha_1 + \mu + \psi)I - \beta_H^* I \\[4pt] \dfrac{d}{dt}R = \alpha_1 I - (\mu + \sigma)R - \beta_H^* R \\[4pt] \dfrac{d}{dt}E = \theta I + \delta C - \nu E \\[4pt] \dfrac{d}{dt}H = \beta_H^*(S+R) - (\alpha_a + \mu + \psi_2)H - \beta_c^* H + (1-r)\gamma_1 C \\[4pt] \dfrac{d}{dt}A = \alpha_a H - (\mu + \psi_2)A - \beta_c^* A + r\gamma_1 C \\[4pt] \dfrac{d}{dt}C = \beta_c^*(H+A) + \beta_H^* I - (\mu + \gamma_1)C \end{cases} \qquad (7.1)$$

where

$$\beta_c^* = \dfrac{\lambda \varepsilon I}{S+I+R+H+A+C} + \rho E, \quad \beta_H^* = \dfrac{\lambda_1 \varepsilon_1 (H+A+gC)}{S+I+R+H+A+C}.$$

In model (7.1), the total population of the individuals is shown by N where $N = S + I + R + H + A + C$, and it is distributed further, mainly in six different classes. These classes include susceptible (S), cryptosporidiosis only I, recovered from cryptosporidiosis only (R), HIV only (H), AIDS only (A), and dual infection (C) individuals. The environmental contamination is shown by E. The susceptible individuals are recruited through Λ. The natural death rate in each class of human is shown by μ, while the disease mortality rate for cryptospirodiosis and HIV-AIDS individuals is, respectively, shown by ψ and ψ_2. The transmission probability is given by λ while the contact rate is shown by ε. Each infected individual due to cryptosporidiosis contributes averagely to the environment given by the parameter θ. At a rate ν, the environment leaves by the crypto. The susceptibility of recovered individuals is given by σ_1 while the recovery rate of crypt is α_1. The contact rate for HIV-AIDS is shown by ε_1, while their transmission probability is shown by λ_1. The causing of co-infection from the environment is shown by the parameter δ, and the modification parameters are g, γ_1 and r. The parameter α_a is the progression from HIV into the AIDS class, and the waning rate of immunity and recovery rate is shown by σ and α_1, respectively. The modification parameter to the environmental treatment is given by ρ. The biological region for the model (7.1) is given by

Fractional Dynamics of HIV-AIDS and Cryptosporidiosis

$$\Omega = \left\{ (S, I, R, H, A, C) \in R_+^6 : N \leq \frac{\Lambda}{\mu}, E \in R_+ : E \leq \frac{\Lambda(\theta + \delta)}{\mu \nu} \right\}$$

where the usual existence, positivity and the other results associated with the epidemic model are holds. The detailed mathematical results associated with this model can be found in [52].

7.3 Basic Concepts of the CF and AB Fractional Derivative

Here, we present the fundamental concepts of the CF and AB fractional order derivative. We have:

Definition 7.1

Consider $\chi \in H^1(a,b)$, $b > a$, $\sigma \in [0,1]$, then the CF fractional derivative [46] is presented as

$$D_t^\sigma(\chi(t)) = \frac{M(\sigma)}{1-\sigma} \int_a^t \chi'(x) \exp\left[-\sigma \frac{t-x}{1-\sigma} \right] dx. \tag{7.2}$$

In Equation (7.2), $M(\sigma)$ indicates the function normality that holds $M(0) = M(1) = 1$ [46] if $\chi \notin H^1(a,b)$ then it can be shown as:

$$D_t^\sigma(\chi(t)) = \frac{\sigma M(\sigma)}{1-\sigma} \int_a^t (\chi(t) - \chi(x)) \exp\left[-\sigma \frac{t-x}{1-\sigma} \right] dx. \tag{7.3}$$

Remark 7.1

If $\xi = \frac{1-\sigma}{\sigma} \in [0,\infty)$, $\sigma = \frac{1}{1+\xi} \in [0,1]$, therefore Equation (7.3) reduces in the following form

$$D_t^\sigma(g(t)) = \frac{M(\sigma)}{\sigma} \int_a^t \chi'(x) \exp\left[-\frac{t-x}{\sigma} \right] dx, M(0) = M(\infty) = 1. \tag{7.4}$$

Moreover,

$$\lim_{\xi \to 0} \frac{1}{\xi} \exp\left[-\frac{t-x}{\xi} \right] = \phi(x-t). \tag{7.5}$$

The corresponding integral was defined in [54].

Definition 7.2

Let $0 < \sigma < 1$, then the integral of fractional order σ for a function $\chi(x)$ is expressed as

$$I_t^\sigma(\chi(t)) = \frac{2(1-\sigma)}{(2-\sigma)\mathcal{M}(\sigma)} g(t) + \frac{2\sigma}{(2-\sigma)\mathcal{M}(\sigma)} \int_0^t \chi(s)ds, t \geq 0. \tag{7.6}$$

Remark 7.2

In Definition 7.2, the remainder of the Caputo-type non-integer order integral of the function having order $0 < \sigma < 1$ is a mean into χ, and its integral is of order 1. Therefore, this requires,

$$\frac{2}{2\mathcal{M}(\sigma) - \sigma \mathcal{M}(\sigma)} = 1, \tag{7.7}$$

which implies $\mathcal{M}(\sigma) = \frac{2}{2-\sigma}, 0 < \sigma < 1$. In view of Equation (7.7), the authors in [54] suggested the new Caputo derivative of order $0 < \sigma < 1$ which is presented as follows:

$$D_t^\sigma(\chi(t)) = \frac{1}{1-\sigma} \int_0^t \chi'(x) \exp\left[-\sigma \frac{t-x}{1-\sigma}\right] dx. \tag{7.8}$$

The definitions of new fractional AB derivatives with non-singular and non-local kernel [53] are presented below.

Definition 7.3

Let $f \in H^1(a,b)$, $b > a$, $\alpha \in [0,1]$ then the new fractional derivatives in the Caputo sense are given below:

$$_a^{ABC}D_t^\sigma(f(t)) = \frac{B(\sigma)}{1-\sigma} \int_a^t f'(\chi) E_\sigma\left[-\sigma \frac{(t-\chi)^\sigma}{1-\sigma}\right] d\chi \tag{7.9}$$

Definition 7.4

Let $f \in H^1(a,b)$, $b > a$, $\sigma \in [0,1]$ and not necessarily differentiable, then the new fractional AB fractional derivative in the Riemann-Liouville sense is given as:

$$_a^{ABR}D_t^\sigma(f(t)) = \frac{B(\sigma)}{1-\sigma} \frac{d}{dt} \int_a^t f(\chi) E_\sigma\left[-\sigma \frac{(t-\chi)^\sigma}{1-\sigma}\right] d\chi. \tag{7.10}$$

Fractional Dynamics of HIV-AIDS and Cryptosporidiosis 173

Definition 7.5

The fractional integral associate to the new fractional derivative with a non-local kernel is defined as:

$$_a^{AB}I_t^\alpha(f(t)) = \frac{1-\sigma}{B(\sigma)}f(t) + \frac{\sigma}{B(\sigma)\Gamma(\sigma)}\int_a^t f(y)(t-y)^{\sigma-1}dy. \quad (7.11)$$

The initial function is recovered when the fractional order turns to zero. Also, when the order turns to 1 we have the classical integral.

Theorem 7.1

Consider f as a continuous function on a closed interval $[a, b]$. Then the following inequality is obtained on $[a, b]$.

$$\|_a^{ABR}D_t^\sigma(f(t))\| < \frac{B(\sigma)}{1-\sigma}\|f(x)\|, \text{ where } \|f(x)\| = \max_{a \le x \le b}|f(x)|. \quad (7.12)$$

Theorem 7.2

Both of (ABR) and (ABC) derivatives satisfy the Lipschitz condition given below:

$$\|_a^{ABR}D_t^\sigma f_1(t) - _a^{ABR}D_t^\sigma f_2(t)\| < K\|f_1(t) - f_2(t)\|, \quad (7.13)$$

also for ABC derivative we have

$$\|_a^{ABC}D_t^\sigma f_1(t) - _a^{ABC}D_t^\sigma f_2(t)\| < K\|f_1(t) - f_2(t)\|. \quad (7.14)$$

Theorem 7.3

The time fractional ordinary differential equation given below

$$_a^{ABC}D_t^\sigma f(t) = s(t), \quad (7.15)$$

has a unique solution with applying the inverse Laplace transform and using the convolution theorem below [53]:

$$f(t) = \frac{1-\sigma}{ABC(\sigma)}s(t) + \frac{\sigma}{ABC(\sigma)\Gamma(\sigma)}\int_a^t s(\xi)(t-\xi)^{\sigma-1}d\xi. \quad (7.16)$$

7.4 HIV-AIDS and Cryptosporidiosis Fractional Model

In the proposed section, we investigate the dynamics of the co-infection model in Equation (7.1), in the sense of the CF fractional derivative. Then, we investigate the model variable existence and their uniqueness. Further, in this section, we proposed the Adams-Bashforth numerical method to investigate their numerical results. Let us consider the model (7.1) in the sense of the CF fractional derivative given by:

$$\begin{cases} {}^{CF}_0 D_t^\sigma S = \Lambda + \sigma_1 R(t) - \mu S(t) - \beta_c^* S(t) - \beta_H^* S(t), \\ {}^{CF}_0 D_t^\sigma I = \beta_c^* S(t) - (\alpha + \mu + \psi) I(t) - \beta_H^* I(t), \\ {}^{CF}_0 D_t^\sigma R = \alpha I(t) - (\mu + \sigma_1) R(t) - \beta_H^* R(t), \\ {}^{CF}_0 D_t^\sigma E = \theta I(t) + \delta C(t) - \nu E(t), \\ {}^{CF}_0 D_t^\sigma H = \beta_H^* (S(t) + R(t)) - (\alpha_a + \mu + \psi_2) H(t) - \beta_c^* H(t) + (1-r)\gamma C(t), \\ {}^{CF}_0 D_t^\sigma A = \alpha_a H(t) - (\mu + \psi_2) A(t) - \beta_c^* A(t) + r\gamma C(t), \\ {}^{CF}_0 D_t^\sigma C = \beta_c^* (H(t) + A(t)) + \beta_H^* I(t) - (\mu + \gamma) C(t), \end{cases} \quad (7.17)$$

where

$$\beta_c^* = \frac{\lambda \varepsilon I}{S + I + R + H + A + C} + \rho E,$$

and

$$\beta_H^* = \frac{\lambda_1 \varepsilon_1 (H + A + gC)}{S + I + R + H + A + C}.$$

7.4.1 Existence and Uniqueness of HIV-AIDS and Cryptosporidiosis Fractional Model

Here, we present the existence of the variables in the co-infection fractional order model in Equation (7.17) by using the concept of the fixed-point theory. Using the concept of [54] on Equation (7.17), we have

$$S(t) - S(0) = {}^{CF}_0 I_t^\sigma \left\{ \Lambda + \sigma_1 R(t) - \mu S(t) - \beta_c^* S(t) - \beta_H^* S(t) \right\},$$

$$I(t) - I(0) = {}^{CF}_0 I_t^\sigma \left\{ \beta_c^* S(t) - (\alpha + \mu + \psi) I(t) - \beta_H^* I(t) \right\}$$

$$R(t) - R(0) = {}^{CF}_0 I_t^\sigma \left\{ \alpha I(t) - (\mu + \sigma_1) R(t) - \beta_H^* R(t) \right\},$$

$$E(t)-E(0) = {}_0^{CF}I_t^\sigma \{\theta I(t)+\delta C(t)-\nu E(t)\},$$

$$H(t)-H(0) = {}_0^{CF}I_t^\sigma \{\beta_H^*(S(t)+R(t))-(\alpha_a+\mu+\psi_2)H(t)-\beta_c^*H(t)+(1-r)\gamma C(t)\},$$

$$A(t)-A(0) = {}_0^{CF}I_t^\sigma \{\alpha_a H(t)-(\mu+\psi_2)A(t)-\beta_c^* A(t)+r\gamma C(t)\},$$

$$C(t)-C_{(0)} = {}_0^{CF}I_t^\sigma \{\beta_c^*(H(t)+A(t))+\beta_H^* I(t)-(\mu+\gamma)C(t)\}. \tag{7.18}$$

Again using [54], we can get

$$S(t)-S(0) = \frac{2(1-\sigma)}{\mathcal{M}(\sigma)(2-\sigma)}\{\Lambda+\sigma_1 R(t)-\mu S(t)-\beta_c^* S(t)-\beta_H^* S(t)\}$$

$$+\frac{2\sigma}{\mathcal{M}(\sigma)(2-\sigma)}\int_0^t \{\Lambda+\sigma R(t)-\mu S(t)-\beta_c^* S(t)-\beta_H^* S(t)\}dy,$$

$$I(t)-I(0) = \frac{2(1-\sigma)}{\mathcal{M}(\sigma)(2-\sigma)}\{\beta_c^* S(t)-(\alpha+\mu+\psi)I(t)-\beta_H^* I(t)\}$$

$$+\frac{2\sigma}{\mathcal{M}(\sigma)(2-\sigma)}\int_0^t \{\beta_c^* S(t)-(\alpha+\mu+\psi)I(t)-\beta_H^* I(t)\}dy,$$

$$R(t)-R(0) = \frac{2(1-\sigma)}{\mathcal{M}(\sigma)(2-\sigma)}\{\alpha I(t)-(\mu+\sigma_1)R(t)-\beta_H^* R(t)\}$$

$$+\frac{2\sigma}{\mathcal{M}(\sigma)(2-\sigma)}\int_0^t \{\alpha I(t)-(\mu+\sigma)R(t)-\beta_H^* R(t)\}dy,$$

$$E(t)-E(0) = \frac{2(1-\sigma)}{\mathcal{M}(\sigma)(2-\sigma)}\{\theta I(t)+\delta C(t)-\nu E(t)\}$$

$$+\frac{2\sigma}{\mathcal{M}(\sigma)(2-\sigma)}\int_0^t \{\theta I(t)+\delta C_{HI}(t)-\nu E(t)\}dy,$$

$$H(t)-H(0) = \frac{2(1-\sigma)}{\mathcal{M}(\sigma)(2-\sigma)}\{\beta_H^*(S(t)+R(t))-(\alpha_a+\mu+\psi_2)H(t)-\beta_c^* H(t)$$

$$+(1-r)\gamma C(t)\}+\frac{2\sigma}{\mathcal{M}(\sigma)(2-\sigma)}\int_0^t \{\beta_H^*(S(t)+R(t))$$

$$-(\alpha_a+\mu+\psi_2)H(t)-\beta_c^* H(t)+(1-r)\gamma C(t)\}dy,$$

$$A(t) - A(0) = \frac{2(1-\sigma)}{M(\sigma)(2-\sigma)} \{\alpha_a H(t) - (\mu + \psi_2)A(t) - \beta_c^* A(t) + r\gamma C(t)\}$$

$$+ \frac{2\sigma}{M(\sigma)(2-\sigma)} \int_0^t \{\alpha_a H(t) - (\mu + \psi_2)A(t) - \beta_c^* A(t) + r\gamma C(t)\} dy,$$

$$C(t) - C(0) = \frac{2(1-\sigma)}{M(\sigma)(2-\sigma)} \{\beta_c^*(H(t) + A(t)) + \beta_H^* I(t) - (\mu + \gamma)C(t)\}$$

$$+ \frac{2\sigma}{M(\sigma)(2-\sigma)} \int_0^t \{\beta_c^*(H(t) + A(t)) + \beta_H^* I(t) - (\mu + \gamma)C(t)\} dy.$$
(7.19)

We interpret for clarity,

$$U_1(t,S) = \Lambda + \sigma_1 R(t) - \mu S(t) - \beta_c^* S(t) - \beta_H^* S(t),$$

$$U_2(t,I) = \beta_c^* S(t) - (\alpha + \mu + \psi)I(t) - \beta_H^* I(t),$$

$$U_3(t,R) = \alpha I(t) - (\mu + \sigma_1)R(t) - \beta_H^* R(t),$$

$$U_4(t,E) = \theta I(t) + \delta C(t) - vE(t),$$

$$U_5(t,H) = \beta_H^*(S(t) + R(t)) - (\alpha_a + \mu + \psi_2)H(t) - \beta_c^* H(t) + (1-r)\gamma C(t),$$

$$U_6(t,A) = \alpha_a H(t) - (\mu + \psi_2)A(t) - \beta_c^* A(t) + r\gamma C(t),$$

$$U_7(t,C) = \beta_c^*(H(t) + A(t)) + \beta_H^* I(t) - (\mu + \gamma)C(t).$$
(7.20)

Theorem 7.4

The kernels U_i for $i = 1,\ldots,7$ holds the Lipchitz condition and contraction if the inequality given below holds

$$0 \leq (\mu + \beta_c^* + \beta_H^*) < 1.$$

Proof

We begin from U_1. Suppose S and S_1 are functions, then the following is presented:

$$\|U_1(t,S) - U_1(t,S_1)\| = \|-(\mu + \beta_c^* + \beta_H^*)(S(t) - S(t_1))\|$$
(7.21)

We have after using the triangular inequality on Equation (7.21),

$$\|U_1(t,S) - U_1(t,S_1)\| \leq \|(\mu)(S(t) - S(t_1))\| + \|(\beta_c^*)(S(t) - S(t_1))\|$$
$$+ \|(\beta_H^*)(S(t) - S(t_1))\| \leq \gamma_1 \|\{S(t) - S(t_1)\}\|.$$
(7.22)

Taking $\gamma_1 = \{\mu + \beta_c^* + \beta_H^*\}$, which implicates

$$\|U_1(t,S) - U_1(t,S_1)\| \leq \gamma_1 \|S(t) - S(t_1)\|. \tag{7.23}$$

This shows the Lipschiz condition for U_1. Further, if the condition holds, $0 \leq (\mu + \beta_c^* + \beta_H^*) < 1$, additionally, then it implies a contraction. For the rest of the variables, one can easily obtain the Lipschiz condition as follows:

$$\|U_2(t,I) - U_2(t,I_1)\| \leq \gamma_2 \|I(t) - I(t_1)\|,$$

$$\|U_3(t,R) - U_3(t,R_1)\| \leq \gamma_3 \|R(t) - R(t_1)\|,$$

$$\|U_4(t,E) - U_4(t,E_1)\| \leq \gamma_4 \|E(t) - E(t_1)\|,$$

$$\|U_5(t,H) - U_5(t,H_1)\| \leq \gamma_5 \|H(t) - H(t_1)\|,$$

$$\|U_6(t,A) - U_6(t,A_1)\| \leq \gamma_6 \|A(t) - A(t_1)\|,$$

$$\|U_7(t,C) - U_7(t,C_1)\| \leq \gamma_7 \|C(t) - C(t_1)\|. \tag{7.24}$$

Equation (7.19) becomes, after the consideration of the previously mentioned kernels,

$$S(t) = S(0) + \frac{2(1-\sigma)}{(2-\sigma)\mathcal{M}(\sigma)} U_1(t,S) + \frac{2\sigma}{(2-\sigma)\mathcal{M}(\sigma)} \int_0^t (U_1(y,S)) dy,$$

$$I(t) = I(0) + \frac{2(1-\sigma)}{(2-\sigma)\mathcal{M}(\sigma)} U_2(t,I) + \frac{2\sigma}{(2-\sigma)\mathcal{M}(\sigma)} \int_0^t (U_2(y,I)) dy,$$

$$R(t) = R(0) + \frac{2(1-\sigma)}{\mathcal{M}(\sigma)(2-\sigma)} U_3(t,R) + \frac{2\sigma}{(2-\sigma)\mathcal{M}(\sigma)} \int_0^t (U_3(y,R)) dy,$$

$$E(t) = E_n(0) + \frac{2(1-\sigma)}{\mathcal{M}(\sigma)(2-\sigma)} U_4(t,E) + \frac{2\sigma}{(2-\sigma)\mathcal{M}(\sigma)} \int_0^t (U_4(y,E)) dy,$$

$$H(t) = H(0) + \frac{2(1-\sigma)}{(2-\sigma)\mathcal{M}(\sigma)} U_5(t,H) + \frac{2\sigma}{(2-\sigma)\mathcal{M}(\sigma)} \int_0^t (U_5(y,H)) dy,$$

$$A(t) = A(0) + \frac{2(1-\sigma)}{(2-\sigma)\mathcal{M}(\sigma)} U_6(t,A) + \frac{2\sigma}{\mathcal{M}(\sigma)(2-\sigma)} \int_0^t (U_6(y,A)) dy,$$

$$C(t) = C(0) + \frac{2(1-\sigma)}{(2-\sigma)\mathcal{M}(\sigma)} U_6(t,C) + \frac{2\sigma}{\mathcal{M}(\sigma)(2-\sigma)} \int_0^t (U_6(y,C)) dy. \tag{7.25}$$

Further, we obtain the recursive formula in the following:

$$S_n(t) = \frac{2(1-\sigma)}{(2-\sigma)\mathcal{M}(\sigma)} U_1(t, S_{n-1}) + \frac{2\sigma}{(2-\sigma)\mathcal{M}(\sigma)} \int_0^t (U_1(y, S_{n-1})) dy,$$

$$I_n(t) = \frac{2(1-\sigma)}{(2-\sigma)\mathcal{M}(\sigma)} U_2(t, I_{n-1}) + \frac{2\sigma}{(2-\sigma)\mathcal{M}(\sigma)} \int_0^t (U_2(y, I_{n-1})) dy,$$

$$R_n(t) = \frac{2(1-\sigma)}{(2-\sigma)\mathcal{M}(\sigma)} U_3(t, R_{n-1}) + \frac{2\sigma}{(2-\sigma)\mathcal{M}(\sigma)} \int_0^t (U_3(y, R_{n-1})) dy,$$

$$E_n(t) = \frac{2(1-\sigma)}{(2-\sigma)\mathcal{M}(\sigma)} U_4(t, E_{n-1}) + \frac{2\sigma}{(2-\sigma)\mathcal{M}(\sigma)} \int_0^t (U_4(y, E_{n-1})) dy,$$

$$H_n(t) = \frac{2(1-\sigma)}{(2-\sigma)\mathcal{M}(\sigma)} U_5(t, H_{n-1}) + \frac{2\sigma}{(2-\sigma)\mathcal{M}(\sigma)} \int_0^t (U_5(y, H_{n-1})) dy,$$

$$A_n(t) = \frac{2(1-\sigma)}{\mathcal{M}(\sigma)(2-\sigma)} U_6(t, A_{n-1}) + \frac{2\sigma}{\mathcal{M}(\sigma)(2-\sigma)} \int_0^t (U_6(y, A_{n-1})) dy,$$

$$C_n(t) = \frac{2(1-\sigma)}{\mathcal{M}(\sigma)(2-\sigma)} U_7(t, C_{n-1}) + \frac{2\sigma}{\mathcal{M}(\sigma)(2-\sigma)} \int_0^t (U_7(y, C_{n-1})) dy. \quad (7.26)$$

The initial conditions are given by

$$\begin{cases} S_0(t) = S(0), \\ I_0(t) = I(0), \\ R_0(t) = R(0), \\ E_0(t) = E(0), \\ H_0(t) = H(0), \\ A_0(t) = A(0), \\ C_0(t) = C(0). \end{cases} \quad (7.27)$$

We have the following expressions for the difference of successive terms:

$$\phi_n(t) = S_n(t) - S_{n-1}(t) = \frac{2(1-\sigma)}{(2-\sigma)\mathcal{M}(\sigma)} (U_1(t, S_{n-1}) - U_1(t, S_{n-2}))$$
$$+ \frac{2\sigma}{(2-\sigma)\mathcal{M}(\sigma)} \int_0^t (U_1(y, S_{n-1}) - U_1(y, S_{n-2})) dy,$$

$$\psi_n(t) = I_n(t) - I_{n-1}(t) = \frac{2(1-\sigma)}{(2-\sigma)\mathcal{M}(\sigma)} (U_2(t, I_{n-1}) - U_2(t, I_{n-2}))$$
$$+ \frac{2\sigma}{(2-\sigma)\mathcal{M}(\sigma)} \int_0^t (U_2(y, I_{n-1}) - U_2(y, I_{n-2})) dy,$$

$$\xi_n(t) = R_n(t) - R_{n-1}(t) = \frac{2(1-\sigma)}{(2-\sigma)\mathcal{M}(\sigma)}(U_3(t,R_{n-1}) - U_3(t,R_{n-2}))$$

$$+ \frac{2\sigma}{(2-\sigma)\mathcal{M}(\sigma)} \int_0^t (U_3(y,R_{n-1}) - U_3(y,R_{n-2})) dy,$$

$$\eta_n(t) = E_n(t) - C_{n-1}(t) = \frac{2(1-\sigma)}{(2-\sigma)\mathcal{M}(\sigma)}(U_4(t,E_{n-1}) - U_4(t,E_{n-2}))$$

$$+ \frac{2\sigma}{(2-\sigma)\mathcal{M}(\sigma)} \int_0^t (K_U(y,E_{n-1}) - U_4(y,E_{n-2})) dy,$$

$$\lambda_n(t) = H_n(t) - H_{n-1}(t) = \frac{2(1-\sigma)}{(2-\sigma)\mathcal{M}(\sigma)}(U_5(t,H_{n-1}) - U_5(t,H_{n-2}))$$

$$+ \frac{2\sigma}{(2-\sigma)\mathcal{M}(\sigma)} \int_0^t (U_5(y,H_{n-1}) - U_5(y,H_{n-2})) dy,$$

$$\zeta_n(t) = A_n(t) - A_{n-1}(t) = \frac{2(1-\sigma)}{(2-\sigma)\mathcal{M}(\sigma)}(U_6(t,A_{n-1}) - U_6(t,A_{n-2}))$$

$$+ \frac{2\sigma}{(2-\sigma)\mathcal{M}(\sigma)} \int_0^t (U_6(y,A_{n-1}) - U_6(y,A_{n-2})) dy,$$

$$\chi_n(t) = C_n(t) - C_{n-1}(t) = \frac{2(1-\sigma)}{(2-\sigma)\mathcal{M}(\sigma)}(U_7(t,C_{n-1}) - U_7(t,C_{n-2}))$$

$$+ \frac{2\sigma}{(2-\sigma)\mathcal{M}(\sigma)} \int_0^t (U_7(y,C_{n-1}) - U_7(y,C_{n-2})) dy.$$

(7.28)

Noticing that

$$\begin{cases} S_n(t) = \sum_{i=1}^n \phi_i(t), \\ I_n(t) = \sum_{i=1}^n \psi_i(t), \\ R_n(t) = \sum_{i=1}^n \xi_i(t), \\ E_n(t) = \sum_{i=1}^n \eta_i(t), \\ H_n(t) = \sum_{i=1}^n \lambda_i(t), \\ A_n(t) = \sum_{i=1}^n \zeta_i(t), \\ C_n(t) = \sum_{i=1}^n \chi_i(t), \end{cases} \qquad (7.29)$$

Step-by-step calculations lead to

$$\begin{aligned}\|\phi_n(t)\| = \|S_n(t) - S_{n-1}(t)\| = \|\frac{2(1-\sigma)}{(2-\sigma)\mathcal{M}(\sigma)}(U_1(t,S_{n-1}) - U_1(t,S_{n-2})) \\ + \frac{2\sigma}{\mathcal{M}(\sigma)(2-\sigma)}\int_0^t (U_1(y,S_{n-1}) - U_1(y,S_{n-2}))dy\|.\end{aligned} \quad (7.30)$$

Equation (7.30) reduces to the following equations after applying the triangular inequality:

$$\begin{aligned}\|S_n(t) - S_{n-1}(t)\| \leq \frac{2(1-\sigma)}{(2-\sigma)\mathcal{M}(\sigma)}\|(U_1(t,S_{n-1}) - U_1(t,S_{n-2}))\| \\ + \frac{2\sigma}{\mathcal{M}(\sigma)(2-\sigma)}\|\int_0^t (U_1(y,S_{n-1}) - U_1(y,S_{n-2}))dy\|.\end{aligned} \quad (7.31)$$

We obtain the following, as the kernel fulfills the Lipchitz condition, so,

$$\begin{aligned}\|S_n(t) - S_{n-1}(t)\| \leq \frac{2(1-\sigma)}{(2-\sigma)\mathcal{M}(\sigma)}\gamma_1\|S_{n-1} - S_{n-2}\| + \frac{2\sigma}{(2-\sigma)\mathcal{M}(\sigma)}\gamma_1 \\ \times \int_0^t \|S_{n-1} - S_{n-2}\|dy.\end{aligned} \quad (7.32)$$

Then, we have

$$\|\phi_n(t)\| \leq \frac{2(1-\sigma)}{(2-\sigma)\mathcal{M}(\sigma)}\gamma_1\|\phi_{n-1}(t)\| + \frac{2\sigma}{(2-\sigma)\mathcal{M}(\sigma)}\gamma_1 \int_0^t \|\phi_{n-1}(y)\|dy. \quad (7.33)$$

In a similar fashion, we obtain the following results:

$$\|\psi_n(t)\| \leq \frac{2(1-\sigma)}{(2-\sigma)\mathcal{M}(\sigma)}\gamma_2\|\psi_{n-1}(t)\| + \frac{2\sigma}{\mathcal{M}(\sigma)(2-\sigma)}\gamma_2 \int_0^t \|\psi_{n-1}(y)\|dy,$$

$$\|\xi_n(t)\| \leq \frac{2(1-\sigma)}{(2-\sigma)\mathcal{M}(\sigma)}\gamma_3\|\xi_{n-1}(t)\| + \frac{2\sigma}{\mathcal{M}(\sigma)(2-\sigma)}\gamma_3 \int_0^t \|\xi_{n-1}(y)\|dy,$$

$$\|\eta_n(t)\| \leq \frac{2(1-\sigma)}{(2-\sigma)\mathcal{M}(\sigma)}\gamma_4\|\eta_{n-1}(t)\| + \frac{2\sigma}{\mathcal{M}(\sigma)(2-\sigma)}\gamma_4 \int_0^t \|\eta_{n-1}(y)\|dy,$$

$$\|\lambda_n(t)\| \leq \frac{2(1-\sigma)}{(2-\sigma)\mathcal{M}(\sigma)}\gamma_5\|\lambda_{n-1}(t)\| + \frac{2\sigma}{\mathcal{M}(\sigma)(2-\sigma)}\gamma_5 \int_0^t \|\lambda_{n-1}(y)\|dy,$$

$$\|\zeta_n(t)\| \leq \frac{2(1-\sigma)}{(2-\sigma)\mathcal{M}(\sigma)}\gamma_6\|\zeta_{n-1}(t)\| + \frac{2\sigma}{\mathcal{M}(\sigma)(2-\sigma)}\gamma_6 \int_0^t \|\zeta_{n-1}(y)\|dy,$$

Fractional Dynamics of HIV-AIDS and Cryptosporidiosis

$$\|\chi_n(t)\| \leq \frac{2(1-\sigma)}{(2-\sigma)\mathcal{M}(\sigma)}\gamma_7\|\chi_{n-1}(t)\| + \frac{2\sigma}{\mathcal{M}(\sigma)(2-\sigma)}\gamma_7\int_0^t\|\chi_{n-1}(y)\|\,dy. \quad (7.34)$$

The following result is established based on the preceding results:

Theorem 7.5

A coupled solution for the HIV-AIDS and cryptosporidiosis fractional model in Equation (7.17) exists if there exists some t_0 such that

$$\frac{2(1-\sigma)}{(2-\sigma)\mathcal{M}(\sigma)}\gamma_1 + \frac{2\sigma}{(2-\sigma)\mathcal{M}(\sigma)}\gamma_1 t_0 < 1$$

Proof

The variables, given by $S(t), I(t), R(t), E(t), H(t), A(t)$, and $C(t)$, satisfy the Lipchitz condition for the kernels and are bounded also. So, by considering equations (7.33) and (7.34), we obtain the following relation by using the recursive method:

$$\|\phi_n(t)\| \leq \|S_n(0)\|\left[\left(\frac{2(1-\sigma)}{\mathcal{M}(\sigma)(2-\sigma)}\gamma_1\right) + \left(\frac{2\sigma}{\mathcal{M}(\sigma)(2-\sigma)}\gamma_1 t\right)\right]^n,$$

$$\|\psi_n(t)\| \leq \|I_n(0)\|\left[\left(\frac{2(1-\sigma)}{\mathcal{M}(\sigma)(2-\sigma)}\gamma_2\right) + \left(\frac{2\sigma}{\mathcal{M}(\sigma)(2-\sigma)}\gamma_2 t\right)\right]^n,$$

$$\|\xi_n(t)\| \leq \|R_n(0)\|\left[\left(\frac{2(1-\sigma)}{\mathcal{M}(\sigma)(2-\sigma)}\gamma_3\right) + \left(\frac{2\sigma}{\mathcal{M}(\sigma)(2-\sigma)}\gamma_3 t\right)\right]^n,$$

$$\|\eta_n(t)\| \leq \|E_n(0)\|\left[\left(\frac{2(1-\sigma)}{2\mathcal{M}(\sigma)-\sigma\mathcal{M}(\sigma)}\gamma_4\right) + \left(\frac{2\sigma}{2\mathcal{M}(\sigma)-\sigma\mathcal{M}(\sigma)}\gamma_4 t\right)\right]^n,$$

$$\|\lambda_n(t)\| \leq \|H_n(0)\|\left[\left(\frac{2(1-\sigma)}{2\mathcal{M}(\sigma)-\sigma\mathcal{M}(\sigma)}\gamma_5\right) + \left(\frac{2\sigma}{2\mathcal{M}(\sigma)-\sigma\mathcal{M}(\sigma)}\gamma_5 t\right)\right]^n,$$

$$\|\zeta_n(t)\| \leq \|A_n(0)\|\left[\left(\frac{2(1-\sigma)}{2\mathcal{M}(\sigma)-\sigma\mathcal{M}(\sigma)}\gamma_6\right) + \left(\frac{2\sigma}{2\mathcal{M}(\sigma)-\sigma\mathcal{M}(\sigma)}\gamma_6 t\right)\right]^n,$$

$$\|\chi_n(t)\| \leq \|C_n(0)\|\left[\left(\frac{2(1-\sigma)}{2\mathcal{M}(\sigma)-\sigma\mathcal{M}(\sigma)}\gamma_7\right) + \left(\frac{2\sigma}{2\mathcal{M}(\sigma)-\sigma\mathcal{M}(\sigma)}\gamma_7 t\right)\right]^n. \quad (7.35)$$

So, we proved the existence of the preceding solutions. Further, we prove that the function is the solution of Equation (7.17), and we set

$$S(t) - S(0) = S_n(t) - B_n(t),$$

$$I(t) - I(0) = I_n(t) - L_n(t),$$

$$R(t) - R(0) = R_n(t) - D_n(t),$$

$$E(t) - E(0) = E(t) - F_n(t),$$

$$H(t) - H(0) = H_n(t) - G_n(t),$$

$$A(t) - A(0) = A_n(t) - J_n(t),$$

$$C(t) - C(0) = C_{HIn}(t) - K_n(t). \tag{7.36}$$

Therefore, we have

$$\|B_n(t)\| = \left\| \frac{2(1-\sigma)}{M(\sigma)(2-\sigma)} (U(t,S) - U(t,S_{n-1})) \right.$$

$$\left. + \frac{2\sigma}{M(\sigma)(2-\sigma)} \int_0^t (U(y,S) - U(y,S_{n-1})) dy \right\|,$$

$$\leq \frac{2(1-\sigma)}{(2-\sigma)M(\sigma)} \|(U(t,S) - U(t,S_{n-1}))\|$$

$$+ \frac{2\sigma}{(2-\sigma)M(\sigma)} \int_0^t \|(U(y,S) - U(y,S_{n-1}))\| dy,$$

$$\leq \frac{2(1-\sigma)}{(2-\sigma)M(\sigma)} \gamma_1 \|S - S_{n-1}\| + \frac{2\sigma}{(2-\sigma)M(\sigma)} \gamma_1 \|S - S_{n-1}\| t. \tag{7.37}$$

The following is obtained by similar fashion:

$$\|B_n(t)\| \leq \left(\frac{2(1-\sigma)}{2M(\sigma) - \sigma M(\sigma)} + \frac{2\sigma}{2M(\sigma) - \sigma M(\sigma)} t \right)^{n+1} \gamma_1^{n+1} a. \tag{7.38}$$

Then at t_0, we have

$$\|B_n(t)\| \leq \left(\frac{2(1-\sigma)}{M(\sigma)(2-\sigma)} + \frac{2\sigma}{M(\sigma)(2-\sigma)} t_0 \right)^{n+1} \gamma_1^{n+1} a. \tag{7.39}$$

Fractional Dynamics of HIV-AIDS and Cryptosporidiosis

Applying a limit on Equation (7.39) as n approaches to ∞, we get $\|B_n(t)\|$ tends to 0. In a similar way, it can be shown that all $\|L_n(t)\|, \|D_n(t)\|, \|F_n(t)\|, \|G_n(t)\|, \|J_n(t)\|, \|K_n(t)\|$ tend to 0. □

For uniqueness of the solutions of model in Equation (7.17), let suppose that there exist another set of solutions of Equation (7.17), which are $S_1(t)$, $I_1(t)$, $R_1(t)$, $E_1(t)$, $H_1(t)$, $A_1(t)$, and $C_1(t)$, then

$$S(t) - S_1(t) = \frac{2(1-\sigma)}{(2-\sigma)\mathcal{M}(\sigma)}(U_1(t,S) - U_1(t,S_1)) \qquad (7.40)$$
$$+ \frac{2\sigma}{(2-\sigma)\mathcal{M}(\sigma)} \int_0^t (U_1(y,S) - U_1(y,S_1))dy.$$

Applying a norm on Equation (7.40), we get

$$\|S(t) - S_1(t)\| \leq \frac{2(1-\sigma)}{(2-\sigma)\mathcal{M}(\sigma)}\|U_1(t,S) - U_1(t,S_1)\| \qquad (7.41)$$
$$+ \frac{2\sigma}{(2-\sigma)\mathcal{M}(\sigma)} \int_0^t \|U_1(y,S) - U_1(y,S_1)\| dy.$$

Using the Lipschitz condition of the kernel leads to

$$\|S(t) - S_1(t)\| \leq \frac{2(1-\sigma)}{(2-\sigma)\mathcal{M}(\sigma)}\gamma_1 \|S(t) - S_1(t)\| \qquad (7.42)$$
$$+ \frac{2\sigma}{(2-\sigma)\mathcal{M}(\sigma)}\gamma_1 t \|S(t) - S_1(t)\|.$$

It gives

$$\|S(t) - S_1(t)\|\left(1 - \frac{2(1-\sigma)}{2\mathcal{M}(\sigma) - \sigma\mathcal{M}(\sigma)}\gamma_1 - \frac{2\sigma}{2\mathcal{M}(\sigma) - \sigma\mathcal{M}(\sigma)}\gamma_1 t\right) \leq 0. \qquad (7.43)$$

Theorem 7.6

A unique system of solutions of the model in Equation (7.17) exists if the following condition holds

$$\left(1 - \frac{2(1-\sigma)}{2\mathcal{M}(\sigma) - \sigma\mathcal{M}(\sigma)}\gamma_1 - \frac{2\sigma}{2\mathcal{M}(\sigma) - \sigma\mathcal{M}(\sigma)}\gamma_1 t\right) > 0.$$

Proof

If the condition in Equation (7.43) holds, then

$$\| S(t) - S_1(t) \| \left(1 - \frac{2(1-\sigma)}{2\mathcal{M}(\sigma) - \sigma \mathcal{M}(\sigma)} \gamma_1 - \frac{2\sigma}{2\mathcal{M}(\sigma) - \sigma \mathcal{M}(\sigma)} \gamma_1 t \right) \leq 0. \quad (7.44)$$

implies that

$$\| S(t) - S_1(t) \| = 0. \quad (7.45)$$

Then, we get

$$S(t) = S_1(t). \quad (7.46)$$

In a similar way, we get

$$\begin{cases} I(t) = I_1(t), \\ R(t) = R_1(t), \\ E(t) = E_1(t), \\ H(t) = H_1(t), \\ A(t) = A_1(t), \\ C(t) = C_1(t). \end{cases} \quad (7.47)$$

which conforms the uniqueness of the system of solutions of model given in Equation (7.17).

Next, we formulate the model in Equation (7.1) in the AB fractional order derivative.

7.5 HIV-AIDS and Cryptosporidiosis with the ABC-Derivative

Here, we write the model in Equation (7.1) in the Atangana-Baleanu in Caputo (ABC) derivative. We have

$${}_0^{ABC}D_t^\sigma S = \Lambda + \sigma_1 R(t) - \mu S(t) - \beta_c^* S(t) - \beta_H^* S(t),$$

$${}_0^{ABC}D_t^\sigma I = \beta_c^* S(t) - (\alpha + \mu + \psi) I(t) - \beta_H^* I(t),$$

$${}_0^{ABC}D_t^\sigma R = \alpha I(t) - (\mu + \sigma_1) R(t) - \beta_H^* R(t),$$

$${}_0^{ABC}D_t^\sigma E = \theta I(t) + \delta C(t) - \nu E(t),$$

$${}_0^{ABC}D_t^\sigma H = \beta_H^*(S(t) + R(t)) - (\alpha_a + \mu + \psi_2) H(t) - \beta_c^* H(t) + (1-r)\gamma C(t),$$

$$^{ABC}_{0}D^\sigma_t A = \alpha_a H(t) - (\mu+\psi_2)A(t) - \beta_c^* A(t) + r\gamma C(t),$$

$$^{ABC}_{0}D^\sigma_t C = \beta_c^*(H(t)+A(t)) + \beta_H^* I(t) - (\mu+\gamma)C_{HI}(t). \tag{7.48}$$

7.5.1 Existence of Solutions for the Co-infection Model

The exact solution of the model in Equation (7.48) cannot be obtained, as the model is non-linear and non-local, and there is no specific method that can handle their exact solution. However, if we can show the existence of the model in Equation (7.48), then we may have the exact solution under some conditions. So, we use here the fixed-point theory to establish the existence results for the model in Equation (7.48). For simplification purposes, we write the model of Equation (7.48) as follows:

$$\begin{cases} ^{ABC}_{0}D^\sigma_t x(t) = F(t,x(t)), \\ x(0) = x_0, \; 0 < t < T < \infty, \end{cases} \tag{7.49}$$

where $x(t)$ is a vector consisting of state variables given by $x(t) = (S, I, R, E_n, H, A, C_{HI})$, and F is a continuous vector function defined as

$$F = \begin{pmatrix} F_1 \\ F_2 \\ F_3 \\ F_4 \\ F_5 \\ F_6 \\ F_7 \end{pmatrix} = \begin{pmatrix} \Lambda + \sigma_1 R(t) - \mu S(t) - \beta_c^* S(t) - \beta_H^* S(t) \\ \beta_c^* S(t) - (\alpha + \mu + \psi)I(t) - \beta_H^* I(t) \\ \alpha I(t) - (\mu+\sigma_1)R(t) - \beta_H^* R(t) \\ \theta I(t) + \delta C(t) - \nu E(t) \\ \beta_H^*(S(t)+R(t)) - (\alpha_a + \mu + \psi_2)H(t) - \beta_c^* H(t) + (1-r)\gamma C(t) \\ \alpha_a H(t) - (\mu+\psi_2)A(t) - \beta_c^* A(t) + r\gamma C(t) \\ \beta_c^*(H(t)+A(t)) + \beta_H^* I(t) - (\mu+\gamma)C(t) \end{pmatrix}.$$

It is easy to prove that the function F satisfies the Lipschitz condition and is given as:

$$\|F(t,x_1(t)) - F(t,x_2(t))\| \le M\|x_1(t) - x_2(t)\|. \tag{7.50}$$

In the following theorem, we prove the existence and uniqueness of the model in Equation (7.48) solution for the AB derivative.

Theorem 7.7

The solution of fractional co-infection model given in Equation (7.48) will be unique under the conditions given below:

$$\frac{(1-\sigma)}{ABC(\sigma)}M + \frac{T^\sigma \sigma}{ABC(\sigma)\Gamma(\sigma)}M < 1. \tag{7.51}$$

Proof

Applying the AB fractional integration to both sides of the system in Equation (7.49), we obtain the following non-linear Volterra integral equation

$$x(t) = x_0 + \frac{1-\sigma}{ABC(\sigma)} F(t, x(t)) + \frac{\sigma}{ABC(\sigma)\Gamma(\sigma)} \int_0^t (t-\eta)^{\sigma-1} F(\eta, x(\eta))d\eta. \quad (7.52)$$

Let $J = (0, T)$ and consider the operator $\Psi : C(J, R^7) \to C(J, R^7)$ defined by

$$\Psi[x(t)] = x_0 + \frac{1-\sigma}{ABC(\sigma)} F(t, x(t)) + \frac{\sigma}{ABC(\sigma)\Gamma(\sigma)} \int_0^t (t-\eta)^{\sigma-1} F(\eta, x(\eta))d\eta. \quad (7.53)$$

Then Equation (7.52) can be written as:

$$x(t) = \Psi[x(t)]. \quad (7.54)$$

The supremum norm on J that is $\|\cdot\|_J$ is given as:

$$\|x(t)\|_J = \sup_{t \in J} \|x(t)\|, \quad x(t) \in C. \quad (7.55)$$

Clearly, $C(J, R^7)$ together with norm $\|\cdot\|_J$ is a Banach space. Applying definition in Equation (7.54) of the operator Ψ, we have

$$\|\Psi[x_1(t)] - \Psi[x_2(t)]\|_J \leq \left\| \frac{(1-\sigma)}{ABC(\sigma)} (F(t, x_1(t)) - F(t, x_2(t)) + \frac{\sigma}{ABC(\sigma)\Gamma(\sigma)} \times \right. \\ \left. \int_0^t (t-\eta)^{\sigma-1} (F(\eta, x_1(\eta)) - F(\eta, x_2(\eta)))d\eta \right\|_J. \quad (7.56)$$

Further making use of triangular inequality and the Lipschitz condition stated in Equation (7.50) and then after simplification, we obtain

$$\|\Psi[x_1(t)] - \Psi[x_1(t)]\|_J \leq \left(\frac{(1-\sigma)\mathcal{M}}{ABC(\sigma)} + \frac{\sigma}{ABC(\sigma)\Gamma(\sigma)} \mathcal{M} T^\sigma \right) \|x_1(t) - x_2(t)\|_J. \quad (7.57)$$

Thus, we finally get

$$\|\Psi[x_1(t)] - \Psi[x_1(t)]\|_J \leq L \|x_1(t) - x_2(t)\|_J, \quad (7.58)$$

where

$$L = \frac{(1-\sigma)\mathcal{M}}{ABC(\sigma)} + \frac{\sigma}{ABC(\sigma)\Gamma(\sigma)} \mathcal{M} T^\sigma.$$

The operator Ψ will be a contraction if the condition in Equation (7.51) fulfilled. Hence, due to the Banach fixed-point theorem, there exists a unique solution of the model in Equation (7.49).

7.6 Numerical Approximations

Here, we investigate the simulations of Equation (7.48) by using the procedure of the Adams-Bashforth method given in [55]. We follow the same procedure as described in [55] to obtain the following. Consider the first equation of the model (7.48) and applying the basic theorem of integration, we obtain

$$S(t) - S_0 = \frac{1-\sigma}{ABC(\sigma)} F_1(S,t) + \frac{\sigma}{ABC(\sigma)\Gamma(\sigma)} \int_0^t f_1(S,\xi)(t-\xi)^{\sigma-1} d\xi.$$

For $t = t_{n+1}$, $n \in 0,1,2,\ldots$ we have

$$S_{n+1} - S_0 = \frac{1-\sigma}{ABC(\sigma)} F_1(S_n,t) + \frac{\sigma}{ABC(\sigma)\Gamma(\sigma)} \int_0^{t_{n+1}} F_1(S,t)(t_{n+1}-t)^{\sigma-1} dt.$$

Further, we obtain the difference between the successive terms as follows:

$$S_{n+1} - S_n = \frac{1-\sigma}{ABC(\sigma)} [F_1(S_n,t) - F_1(S_{n-1},t)] + \frac{\sigma}{ABC(\sigma)\Gamma(\sigma)} (I_{\sigma,1} - I_{\sigma,2}), \quad (7.59)$$

where $I_{\sigma,1}$ and $I_{\sigma,2}$ are given by

$$I_{\sigma,1} = \int_0^{t_{n+1}} F_1(S,t)(t_{n+1}-t)^{\sigma-1} dt,$$

$$I_{\sigma,2} = \int_0^{t_n} F_1(S,t)(t_{n+1}-t)^{\sigma-1} dt.$$

The function $F_1(S,t)$ is further approximated by the two points polynomial interpolations, given by

$$L(t) (\simeq F_1(S,t)) = \frac{t-t_{n-1}}{h} F(t_n, y(t_n)) - \frac{t-t_n}{h} F(t_{n-1}, y(t_{n-1})), \quad (7.60)$$

where $h = t_n - t_{n-1}$. We then have the following

$$I_{\sigma,1} = \frac{F_1(S_n,t)}{h} \left[-\frac{2h}{\sigma} t_{n+1}^\sigma + \frac{t_{n+1}^{\sigma+1}}{\sigma+1} \right] + \frac{F_1(S_{n-1},t)}{h} \left[\frac{h}{\sigma} t_{n+1}^\sigma - \frac{t_{n+1}^{\sigma+1}}{\sigma+1} \right]. \quad (7.61)$$

Similarly,

$$I_{\sigma,2} = \frac{F_1(S_n,t)}{h} \left[-\frac{h}{\sigma} t_n^\sigma + \frac{t_n^{\sigma+1}}{\sigma+1} \right] + \frac{F_1(S_n,t)}{h(\sigma+1)} t_n^{\sigma+1}. \quad (7.62)$$

On putting equations (7.61) and (7.62) in Equation (7.59), we obtained the approximate solution given below

$$S_{n+1} = S_n + F_1(S_n,t)\left[\frac{1-\sigma}{ABC(\sigma)} + \frac{\sigma}{ABC(\sigma)h\Gamma(\sigma)}\left(-\frac{2h}{\sigma}t_{n+1}^\sigma + \frac{t_{n+1}^{\sigma+1}}{\sigma+1} + \frac{h}{\sigma}t_n^\sigma - \frac{t_n^{\sigma+1}}{\sigma+1}\right)\right]$$

$$+ F_1(S_{n-1},t)\left[\frac{\sigma-1}{ABC(\sigma)} + \frac{\sigma}{ABC(\sigma)h\Gamma(\sigma)}\left(\frac{h}{\sigma}t_{n+1}^\sigma - \frac{t_{n+1}^{\sigma+1}}{\sigma+1} - \frac{t_n^{\sigma+1}}{\sigma+1}\right)\right]. \quad (7.63)$$

Following the same method, we can obtain the following expression for the variables in the systems:

$$I_{n+1} = I_n + F_2(S_n,t)\left[\frac{1-\sigma}{ABC(\sigma)} + \frac{\sigma}{ABC(\sigma)h\Gamma(\sigma)}\left(-\frac{2h}{\sigma}t_{n+1}^\sigma + \frac{t_{n+1}^{\sigma+1}}{\sigma+1} + \frac{h}{\sigma}t_n^\sigma - \frac{t_n^{\sigma+1}}{\sigma+1}\right)\right]$$

$$+ F_2(S_{n-1},t)\left[\frac{\sigma-1}{ABC(\sigma)} + \frac{\sigma}{ABC(\sigma)h\Gamma(\sigma)}\left(\frac{h}{\sigma}t_{n+1}^\sigma - \frac{t_{n+1}^{\sigma+1}}{\sigma+1} - \frac{t_n^{\sigma+1}}{\sigma+1}\right)\right]. \quad (7.64)$$

$$R_{n+1} = R_n + F_3(S_n,t)\left[\frac{1-\sigma}{ABC(\sigma)} + \frac{\sigma}{ABC(\sigma)h\Gamma(\sigma)}\left(-\frac{2h}{\sigma}t_{n+1}^\sigma + \frac{t_{n+1}^{\sigma+1}}{\sigma+1} + \frac{h}{\sigma}t_n^\sigma - \frac{t_n^{\sigma+1}}{\sigma+1}\right)\right]$$

$$+ F_3(S_{n-1},t)\left[\frac{\sigma-1}{ABC(\sigma)} + \frac{\sigma}{ABC(\sigma)h\Gamma(\sigma)}\left(\frac{h}{\sigma}t_{n+1}^\sigma - \frac{t_{n+1}^{\sigma+1}}{\sigma+1} - \frac{t_n^{\sigma+1}}{\sigma+1}\right)\right]. \quad (7.65)$$

$$E_{n+1} = E_n + F_4(S_n,t)\left[\frac{1-\sigma}{ABC(\sigma)} + \frac{\sigma}{ABC(\sigma)h\Gamma(\sigma)}\left(-\frac{2h}{\sigma}t_{n+1}^\sigma + \frac{t_{n+1}^{\sigma+1}}{\sigma+1} + \frac{h}{\sigma}t_n^\sigma - \frac{t_n^{\sigma+1}}{\sigma+1}\right)\right]$$

$$+ F_4(S_{n-1},t)\left[\frac{\sigma-1}{ABC(\sigma)} + \frac{\sigma}{ABC(\sigma)h\Gamma(\sigma)}\left(\frac{h}{\sigma}t_{n+1}^\sigma - \frac{t_{n+1}^{\sigma+1}}{\sigma+1} - \frac{t_n^{\sigma+1}}{\sigma+1}\right)\right]. \quad (7.66)$$

$$H_{n+1} = H_n + F_5(S_n,t)\left[\frac{1-\sigma}{ABC(\sigma)} + \frac{\sigma}{ABC(\sigma)h\Gamma(\sigma)}\left(-\frac{2h}{\sigma}t_{n+1}^\sigma + \frac{t_{n+1}^{\sigma+1}}{\sigma+1} + \frac{h}{\sigma}t_n^\sigma - \frac{t_n^{\sigma+1}}{\sigma+1}\right)\right]$$

$$+ F_5(S_{n-1},t)\left[\frac{\sigma-1}{ABC(\sigma)} + \frac{\sigma}{ABC(\sigma)h\Gamma(\sigma)}\left(\frac{h}{\sigma}t_{n+1}^\sigma - \frac{t_{n+1}^{\sigma+1}}{\sigma+1} - \frac{t_n^{\sigma+1}}{\sigma+1}\right)\right]. \quad (7.67)$$

Fractional Dynamics of HIV-AIDS and Cryptosporidiosis

$$A_{n+1} = A_n + F_6(S_n,t) \left[\frac{1-\sigma}{ABC(\sigma)} + \frac{\sigma}{ABC(\sigma)h\Gamma(\sigma)} \left(-\frac{2h}{\sigma} t_{n+1}^\sigma + \frac{t_{n+1}^{\sigma+1}}{\sigma+1} + \frac{h}{\sigma} t_n^\sigma - \frac{t_n^{\sigma+1}}{\sigma+1} \right) \right]$$

$$+ F_6(S_{n-1},t) \left[\frac{\sigma-1}{ABC(\sigma)} + \frac{\sigma}{ABC(\sigma)h\Gamma(\sigma)} \left(\frac{h}{\sigma} t_{n+1}^\sigma - \frac{t_{n+1}^{\sigma+1}}{\sigma+1} - \frac{t_n^{\sigma+1}}{\sigma+1} \right) \right]. \quad (7.68)$$

$$C_{n+1} = C_n + F_7(S_n,t) \left[\frac{1-\sigma}{ABC(\sigma)} + \frac{\sigma}{ABC(\sigma)h\Gamma(\sigma)} \left(-\frac{2h}{\sigma} t_{n+1}^\sigma + \frac{t_{n+1}^{\sigma+1}}{\sigma+1} + \frac{h}{\sigma} t_n^\sigma - \frac{t_n^{\sigma+1}}{\sigma+1} \right) \right]$$

$$+ F_7(S_{n-1},t) \left[\frac{\sigma-1}{ABC(\sigma)} + \frac{\sigma}{ABC(\sigma)h\Gamma(\sigma)} \left(\frac{h}{\sigma} t_{n+1}^\sigma - \frac{t_{n+1}^{\sigma+1}}{\sigma+1} - \frac{t_n^{\sigma+1}}{\sigma+1} \right) \right]. \quad (7.69)$$

For Equations (7.63–7.69), we can easily obtain the stability analysis as well its convergence. For this purpose, we refer the readers to see [55], and the results therein.

7.7 Numerical Solution

The numerical scheme previously presented is successfully used to obtain the numerical solution of the model with the CF derivative and the mode with the AB derivative. The graphical results for both the systems of fractional orders are given in Figures 7.1 through 7.11. The parameter values considered in this simulation are $\Lambda = 0.5$, $\sigma_1 = 0.002$, $\psi = 0.0006$, $g = 0.2$, $\mu = 0.0001$, $\theta = 0.02$, $v = 0.385$, $\lambda_1 = 0.23$, $\rho = 0.00321$, $\delta = 0.2$, $r = 0.002$, $\psi_2 = 0.006$, $\gamma_1 = 0.04$, $\alpha_a = 0.006$, $\alpha = 0.004$, $\lambda = 0.04$, $\varepsilon = 0.04$, and $\varepsilon_1 = 0.24$. For the numerical results, we considered different values of the fractional order parameters σ. Figures 7.1 through 7.4 show the graphical results for the AB fractional derivative, while Figures 7.5 through 7.8 show the graphical results for the CF derivative. A comparison for different fractional values of the parameter, comparison results, are presented in Figures 7.9 through 7.11. In these graphical results, we consider the values of the fractional order parameter $\sigma = 1, 0.95, 0.90, 0.85$. It can be seen from Figures 7.9 through 7.11 that the results of the AB fractional derivative for infective compartment are rapidly decreasing more than that of the CF fractional order derivative.

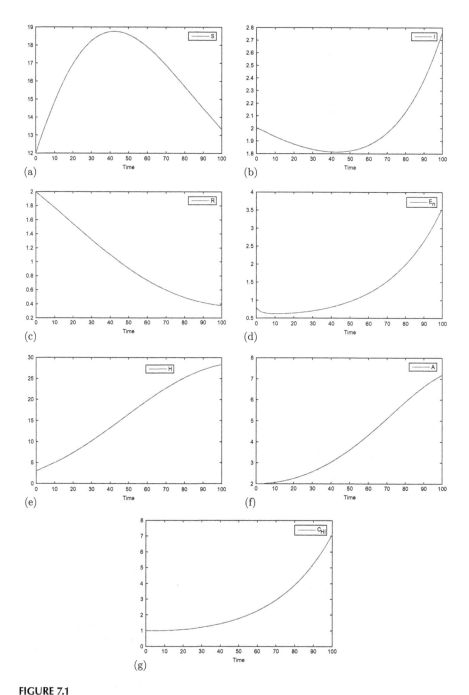

FIGURE 7.1
The behaviour of the model variables for $\sigma = 1$: (a) susceptible individuals, (b) infected with cryptosporidiosis only, (c) recovered from cryptosporidiosis, (d) environmental contamination, (e) HIV only, (f) with AIDS only, and (g) co-infected individuals.

Fractional Dynamics of HIV-AIDS and Cryptosporidiosis 191

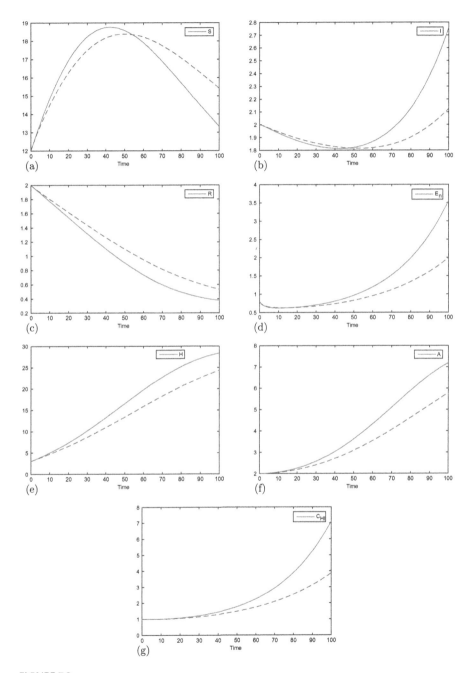

FIGURE 7.2
The behaviour of the model variables for $\sigma = 0.95$: (a) susceptible individuals, (b) infected with cryptosporidiosis only, (c) recovered from cryptosporidiosis, (d) environmental contamination, (e) HIV only, (f) with AIDS only, and (g) co-infected individuals.

192 *Fractional Calculus in Medical and Health Science*

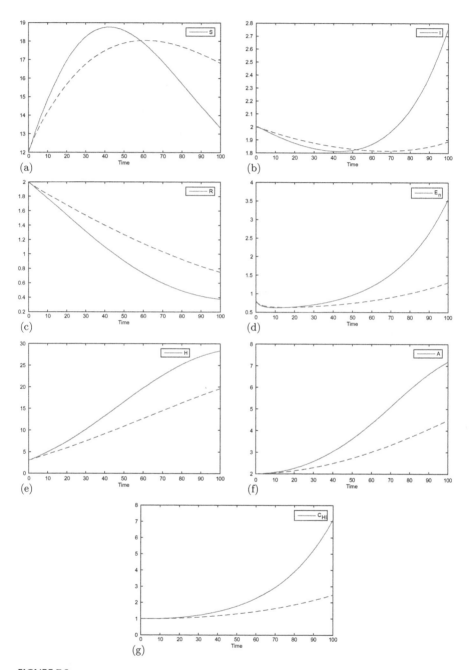

FIGURE 7.3
The behaviour of the model variables for $\sigma = 0.90$: (a) susceptible individuals, (b) infected with cryptosporidiosis only, (c) recovered from cryptosporidiosis, (d) environmental contamination, (e) HIV only, (f) with AIDS only, and (g) co-infected individuals.

Fractional Dynamics of HIV-AIDS and Cryptosporidiosis 193

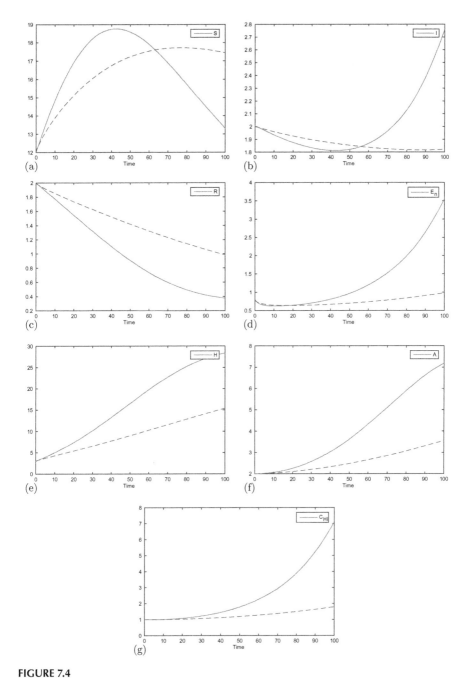

FIGURE 7.4
The behaviour of the model variables for $\sigma = 0.85$: (a) susceptible individuals, (b) infected with cryptosporidiosis only, (c) recovered from cryptosporidiosis, (d) environmental contamination, (e) HIV only, (f) with AIDS only, and (g) co-infected individuals.

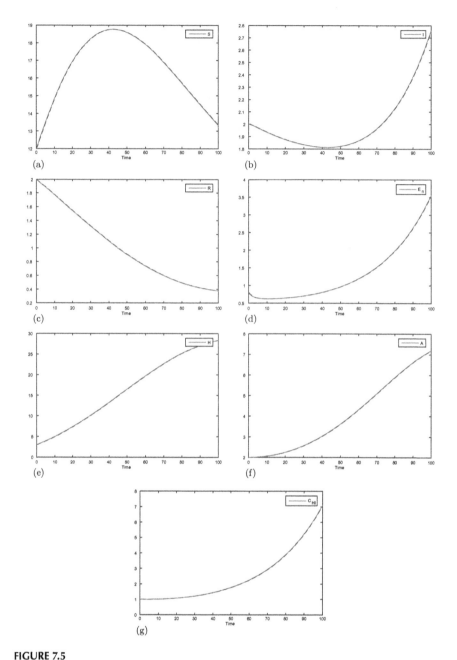

FIGURE 7.5
The behaviour of the model variables for the CF derivative when $\sigma = 1$: (a) susceptible individuals, (b) infected with cryptosporidiosis only, (c) recovered from cryptosporidiosis, (d) environmental contamination, (e) HIV only, (f) with AIDS only, and (g) co-infected individuals.

Fractional Dynamics of HIV-AIDS and Cryptosporidiosis

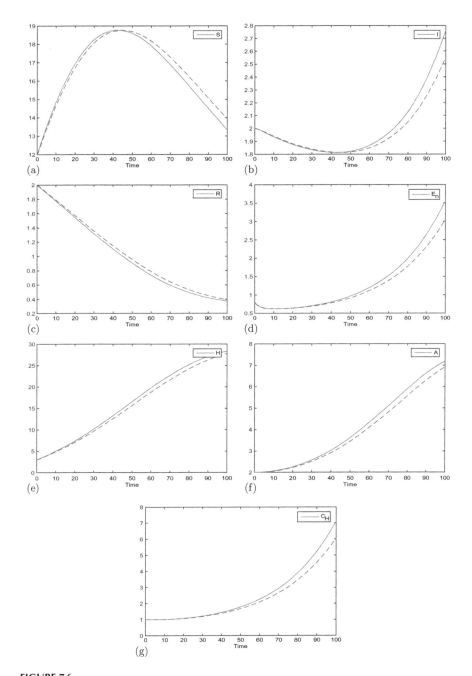

FIGURE 7.6
The behaviour of the model variables for the CF derivative when $\sigma = 0.95$: (a) susceptible individuals, (b) infected with cryptosporidiosis only, (c) recovered from cryptosporidiosis, (d) environmental contamination, (e) HIV only, (f) with AIDS only, and (g) co-infected individuals.

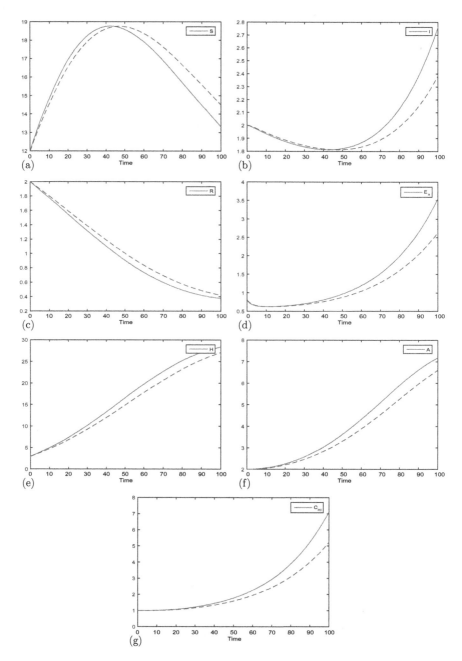

FIGURE 7.7
The behaviour of the model variables for the CF derivative when $\sigma = 0.90$: (a) susceptible individuals, (b) infected with cryptosporidiosis only, (c) recovered from cryptosporidiosis, (d) environmental contamination, (e) HIV only, (f) with AIDS only, and (g) co-infected individuals.

Fractional Dynamics of HIV-AIDS and Cryptosporidiosis 197

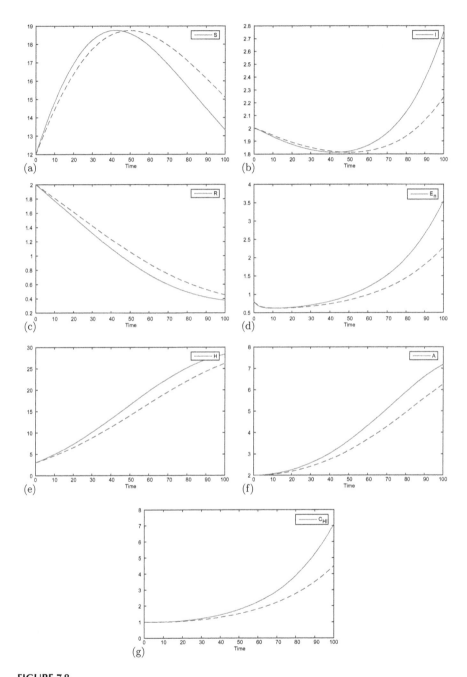

FIGURE 7.8
The behaviour of the model variables for the CF derivative when $\sigma = 0.85$: (a) susceptible individuals, (b) infected with cryptosporidiosis only, (c) recovered from cryptosporidiosis, (d) environmental contamination, (e) HIV only, (f) with AIDS only, and (g) co-infected individuals.

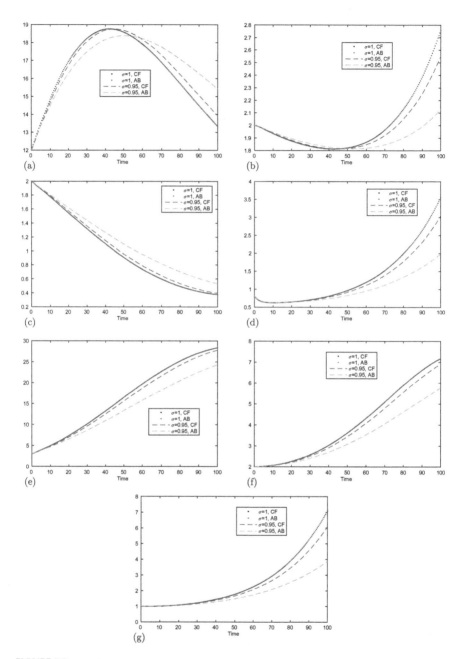

FIGURE 7.9
Comparison of CF and AB derivative when $\sigma = 0.95$: (a) susceptible individuals, (b) infected with cryptosporidiosis only, (c) recovered from cryptosporidiosis, (d) environmental contamination, (e) HIV only, (f) with AIDS only, and (g) co-infected individuals.

Fractional Dynamics of HIV-AIDS and Cryptosporidiosis 199

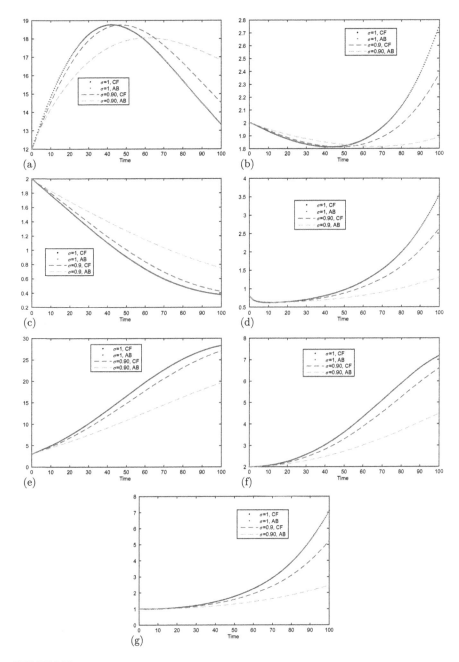

FIGURE 7.10
Comparison of CF and AB derivative when $\sigma = 0.90$: (a) susceptible individuals, (b) infected with cryptosporidiosis only, (c) recovered from cryptosporidiosis, (d) environmental contamination, (e) HIV only, (f) with AIDS only, and (g) co-infected individuals.

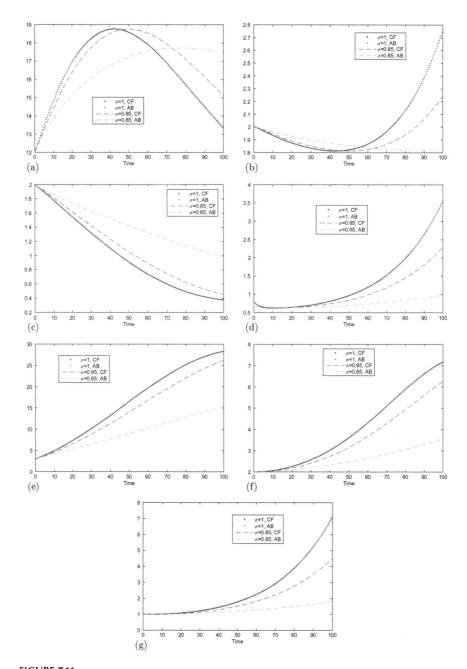

FIGURE 7.11
Comparison of CF and AB derivative when $\sigma = 0.85$: (a) susceptible individuals, (b) infected with cryptosporidiosis only, (c) recovered from cryptosporidiosis, (d) environmental contamination, (e) HIV only, (f) with AIDS only, and (g) co-infected individuals.

7.8 Model with Stochastic Approach: Application of Lognormal Distribution

For simplicity, many models explaining the spread of infectious diseases suggested that coefficients or parameters used in the models are constant. However, the real-world situation suggests that those parameters are not constants, but rather they varied within a range. In order to capture the effect of such parameters, some important works suggested the use of stochastic modelling, where the deterministic model is converted to a stochastic model. The conversion imposes the replacement of constant parameters with distributions. One of the most used distributions is perhaps the normal distribution. Nevertheless, all spread of disease does not follow the normal distribution or maybe the coefficient used in these models. In this section, for comparison purposes, we suggest analysing the considered model using the lognormal distribution for each parameter. To achieve this, we replace model (7.18) parameters by the following way,

$$(b_j), j = 1,2,3,\ldots b_j \sim \log N(m_j, \delta_j^2).$$

Every parameter of this model will be reformulated as:

$$b_j = m_j + v_j X_j, \tag{7.70}$$

where X is the lognormal distribution given as

$$X_j = \exp(m_j + \delta_j Z_j). \tag{7.71}$$

The mean and standard deviation are, respectively, shown by m_j, δ_j of the variable's natural logarithm, and Z_j is a normally distributed random variable with expected value 0 and variance 1. We propose a novel approach to the model by considering the coefficients as random variables that can show the variations in the results. The coefficients of the numerical systems of equations by nature are random, so we can make a system of random differential equations by transferring the condiments of the model (7.1) to random variables. Therefore, the following set of differential equations is obtained with the set of random coefficients:

$$\Lambda \sim N(a_1, s_1^2), \sigma_1 \sim N(a_2, s_2^2), \psi \sim N(a_3, s_3^2),$$

$$g \sim N(a_4, s_4^2), \mu \sim N(a_5, s_5^2), \theta \sim N(a_6, s_6^2),$$

$$v \sim N(a_7, s_7^2), \lambda_1 \sim N(a_8, s_8^2), \rho \sim N(a_9, s_9^2), \delta \sim N(a_{10}, s_{10}^2), r \sim N(a_{11}, s_{11}^2),$$

$$\psi_2 \sim N(a_{12}, s_{12}^2), \gamma_1 \sim N(a_{13}, s_{13}^2), \alpha_a \sim N(a_{14}, s_{14}^2), \alpha \sim N(a_{15}, s_{15}^2),$$

$$\lambda \sim N(a_{16}, s_{16}^2), \varepsilon \sim N(a_{17}, s_{17}^2), \varepsilon_1 \sim N(a_{18}, s_{18}^2). \tag{7.72}$$

where $a_i, i = (\overline{1,18})$ is the means, and $s_i, i = (\overline{1,18})$ is the standard deviations of the normal distributions. Various factors are involved in the values of the coefficients that can affect it. The random variables given often have a normal distribution because it is the expected as a sum of independent quantities. A normal distribution can be used when the random variables are large in number and whose distributions are not known; this fact comes from the central limit theorem. So, here, we are considering a large number of random variables, and thus it is reasonable to use the normal distribution. We will choose the mean values of the distributions according to coefficients of numerical values, and the standard deviations of the distributions will be 5% of the same value.

$N(a, b^2)$ is the distributed random variable X, and it can be shown by $X = a + b\chi$, where the expression shows $\chi \sim N(0,1)$ the standard normally distributed random variable. If it follows from this property, we can express the random variables as standard normally distributed random variables, given by

$$\Lambda = a_1 + s_1\chi_1, \sigma_1 = a_2 + s_2\chi_2, \psi = a_3 + s_3\chi_3, g = a_4 + s_4\chi_4,$$

$$\mu = a_5 + s_5\chi_5, \theta = a_6 + s_6\chi_6,$$

$$v = a_7 + s_7\chi_7, \lambda_1 = a_8 + s_8\chi_8, \rho = a_9 + s_9\chi_9, \delta = a_{10} + s_{10}\chi_{10}, r = a_{11} + s_{11}\chi_{11},$$

$$\psi_2 = a_{12} + s_{12}\chi_{12}, \gamma_1 = a_{13} + s_{13}\chi_{13}, \alpha_a = a_{14} + s_{14}\chi_{14}, \alpha = a_{15} + s_{15}\chi_{15},$$

$$\lambda = a_{16} + s_{16}\chi_{16}, \varepsilon = a_{17} + s_{17}\chi_{17}, \varepsilon_1 = a_{18} + s_{18}\chi_{18}, \tag{7.73}$$

for independent random variables $\chi_i, i = (\overline{1,18})$ with distribution $N(0,1)$. Setting the appropriate values of $a_i, i = (\overline{1,18})$ and $s_i, i = (\overline{1,18})$ yields

$$\Lambda = 0.5 + 0.9\chi_1, \sigma_1 = 0.002 + 0.9\chi_2, \psi = 0.0006 + 0.9\chi_3,$$

$$g = 0.2 + 0.9\chi_4, \mu = 0.00016 + 0.9\chi_5,$$

$$\theta = 0.02 + 0.9\chi_6, v = 0.385 + 0.9\chi_7, \lambda_1 = 0.23 + 0.9\chi_8,$$

$$\rho = 0.00321 + 0.9\chi_9, \delta = 0.2 + 0.9\chi_{10},$$

$r = 0.002 + 0.9\chi_{11}, \psi_2 = 0.006 + 0.9\chi_{12}, \gamma_1 = 0.04 + 0.9\chi_{13}, \alpha_a = 0.006 + 0.9\chi_{14},$

$\alpha_1 = 0.004 + 0.9\chi_{15}, \lambda = 0.04 + 0.9\chi_{16}, \varepsilon = 0.04 + 0.9\chi_{17}, \varepsilon_1 = 0.24 + 0.9\chi_{18},$ (7.74)

with initial conditions,

$$S(0) = 12, I(0) = 2, R(0) = 2, E_n(0) = 0.8, H(0) = 3, A(0) = 2, C_{HI} = 1.$$

We replace the model parameters in Equation (7.1) by the new set of random variables, and the following new system is obtained:

$$\begin{cases} \dfrac{d}{dt}S = 0.5 + 0.9\chi_1 + (0.002 + 0.9\chi_2)R - (0.00016 + 0.9\chi_5)S - \beta_c^* S - \beta_H^* S \\[4pt] \dfrac{d}{dt}I = \beta_c^* S - (0.004 + 0.9\chi_{15} + 0.00016 + 0.9\chi_5 + 0.0006 + 0.9\chi_3)I - \beta_H^* I \\[4pt] \dfrac{d}{dt}R = (0.004 + 0.9\chi_{15})I - (0.00016 + 0.9\chi_5 + 0.002 + 0.9\chi_2)R - \beta_H^* R \\[4pt] \dfrac{d}{dt}E = (0.02 + 0.9\chi_6)I + (0.2 + 0.9\chi_{10})C - (0.385 + 0.9\chi_7)E \\[4pt] \dfrac{d}{dt}H = \beta_H^*(S+R) - (0.006 + 0.9\chi_{14} + 0.00016 + 0.9\chi_5 + 0.006 + 0.9\chi_{12})H \\[4pt] \qquad -\beta_c^* H + (1 - 0.002 + 0.9\chi_{11})(0.04 + 0.9\chi_{13})C \\[4pt] \dfrac{d}{dt}A = (0.006 + 0.9\chi_{14})H - (0.00016 + 0.9\chi_5 + 0.006 + 0.9\chi_{12})A \\[4pt] \qquad -\beta_c^* A + (0.002 + 0.9\chi_{11})(0.04 + 0.9\chi_{13})C \\[4pt] \dfrac{d}{dt}C = \beta_c^*(H + A) + \beta_H^* I - (0.00016 + 0.9\chi_5 + 0.04 + 0.9\chi_{13})C \end{cases}$$ (7.75)

where

$$\beta_c^* = \dfrac{(0.04 + 0.9\chi_{16})(0.04 + 0.9\chi_{17})I}{S + I + R + H + A + C} + (0.00321 + 0.9\chi_9)E,$$

and $\beta_H^* = \dfrac{(0.24 + 0.9\chi_{18})(0.23 + 0.9\chi_8)(H + A + (0.2 + 0.9\chi_4)C)}{S + I + R + H + A + C}.$

Next, we present the numerical solution of the model with stochastic approach.

7.8.1 Numerical Solution of Random Model for HIV-AIDS and Cryptosporidiosis

In this subsection, we obtain the numerical results of the model in Equation (7.75), by using the Adam-Bashforth method of ordinary differential equation, and the following solution is obtained:

$$S_{n+1} = S_n + \frac{3h}{2}\Big[0.5 + 0.9\chi_1 + (0.002 + 0.9\chi_2)R_n - (0.00016 + 0.9\chi_5)S_n - \beta_c^* S_n - \beta_H^* S_n\Big],$$

$$I_{n+1} = I_n + \frac{3h}{2}\Big[\beta_c^* S_n - (0.004 + 0.9\chi_{15} + 0.00016 + 0.9\chi_5 + 0.0006 + 0.9\chi_3)I_n - \beta_H^* I_n\Big],$$

$$R_{n+1} = R_n + \frac{3h}{2}\Big[(0.004 + 0.9\chi_{15})I_n - (0.00016 + 0.9\chi_5 + 0.002 + 0.9\chi_2)R_n - \beta_H^* R_n\Big],$$

$$E_{n+1} = E_n + \frac{3h}{2}\Big[(0.02 + 0.9\chi_6)I_n + (0.2 + 0.9\chi_{10})C_n - (0.385 + 0.9\chi_7)E_n\Big],$$

$$H_{n+1} = H_n + \frac{3h}{2}\Big[\beta_H^*(S_n + R_n) - (0.006 + 0.9\chi_{14} + 0.00016 + 0.9\chi_5 + 0.006 + 0.9\chi_{12})H_n$$
$$- \beta_c^* H_n + (1 - 0.002 + 0.9\chi_{11})(0.04 + 0.9\chi_{13})C_n\Big],$$

$$A_{n+1} = A_n + \frac{3h}{2}\Big[(0.006 + 0.9\chi_{14})H_n - (0.00016 + 0.9\chi_5 + 0.006 + 0.9\chi_{12})A_n$$
$$- \beta_c^* A_n + (0.002 + 0.9\chi_{11})(0.04 + 0.9\chi_{13})C_n\Big],$$

$$C_{n+1} = C_n + \frac{3h}{2}\Big[\beta_c^*(H_n + A_n) + \beta_H^* I_n - (0.00016 + 0.9\chi_5 + 0.04 + 0.9\chi_{13})C_n\Big]. \quad (7.76)$$

And the corresponding graphical result for the system in Equation (7.76) with the same initial conditions are considered in the model in equations (7.17) and (7.48), except the parameters. The parameters used in the simulation of the model with a stochastic approach and the final results by the Adam-Bashforth (equation 7.76) are shown in Figure 7.12. From Figure 7.12, we can observe that the model with the stochastic approach has the same behaviour with the model in equations (7.17) and (7.48) (see Figures 7.1 and 7.5).

Fractional Dynamics of HIV-AIDS and Cryptosporidiosis 205

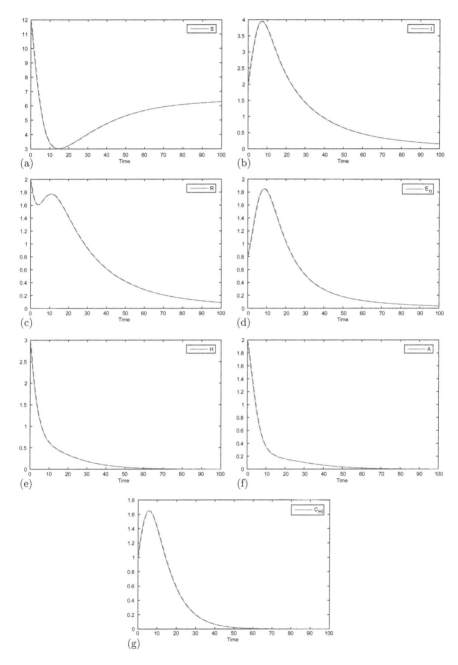

FIGURE 7.12
Numerical solution of a random model in Equation (7.75): (a) susceptible individuals, (b) infected with cryptosporidiosis only, (c) recovered from cryptosporidiosis, (d) environmental contamination, (e) HIV only, (f) with AIDS only, and (g) co-infected individuals.

7.9 Conclusion

Due to the complexities and the fatality of the infectious disease known as HIV/AIDS and crytospirodiosis, researchers in different fields are determined to understand, analyse and predict the behaviour of such diseases in time. Using the classical differential operator, the model will eventually lead to a Markovian process, meaning one will be able to predict the future behaviour only with the initial condition at a given time. This process is not observed in real-world situations; therefore another concept of differentiation must be considered. To capture some complexities of the real-world problem using mathematical formulas, the concept of a non-local differential operator is adopted. However, each concept or formula has its own limitation and advantage; thus, to capture the exponential decay law observed in the real-world problem, we adopt a fractional differential operator with an exponential waiting distribution, which is the CF operator. To capture a crossover behaviur observed in the dynamical system, we adopt the fractional differential operator possessing such properties, which is the AB fractional differential operator. Now assuming that the observed facts present Markovian behaviour with some randomness, we capture such behaviour using lognormal distribution. To be mathematically consistent, we presented an analysis underpinning the conditions of existence and uniqueness of exact solutions within the framework of a fixed-point theorem. Numerical representations of the model with lognormal distribution shows that this distribution could be adequate to model some real-world problems very efficiently, as some real-world input parameters do not always follow the normal distribution.

References

1. F. D. Chierico, M. Onori, S. D. Bella, E. Bordi, N. Petrosillo, D. Menichella, S. M. Caccio, F. Callea, and L. Putignani. Cases of cryptosporidiosis co-infections in AIDS patients: A correlation between clinical presentation and GP60 subgenotype lineages from aged formalin-fixed stool samples. *Annals of Tropical Medicine and Parasitology,* 105 (5) (2011), 339–349.
2. S. T. Ogunlade, K. O. Okosun, R. S. Lebelo, and M. Mukamuri. Optimal control analysis of cryptosporidiosis disease. *Global Journal of Pure and Applied Mathematics,* 12 (6) (2016), 4959–4989.
3. N. Egloff, T. Oehler, M. Rossi, X. M. Nguyen, and H. Furrer. Chronic watery diarrhea due to co-infection with Cryptosporidium spp and Cyclospora cayetanensis in a Swiss AIDS patient traveling in Thailand. *Journal of Travel Medicine,* 8 (3) (2001), 143–145.
4. S. M. Mor and S. Tzipori. Cryptosporidiosis in children in Sub-Saharan Africa: A lingering challenge. *Clinical Infectious Diseases,* 47 (2008), 915.

5. L.-G. Tian, T.-P. Wang, J.-X. Chen, et al. Co-infection of HIV and parasites in China: Results from an epidemiological survey in rural areas of Fuyang city, Anhui province, China. *Frontiers of Medicine in China*, 4 (2) (2010), 192–198.
6. Z. Bentwich, Z. Weisman, C. Moroz, S. Bar-Yehuda, and A. Kalinkovich. Immune dysregulation in Ethiopian immigrants in Israel: Relevance to helminth infections? *Clinical & Experimental Immunology*, 103 (2) (1996), 239–243.
7. R. Gopinath, M. Ostrowski, S. J. Justement, A. S. Fauci, and T. B. Nutman. Infections increase susceptibility to human immunodeficiency virus infection in peripheral blood mononuclear cells in vitro. *Journal of Infectious Diseases*, 182 (6) (2000), 1804–1808.
8. A. Chandora, P. K. Khatri, S. Meena, A. Bora, L. Rathore, V. Maurya, and K. Sirvi. Co-infection with cryptosporidium, isospora and giardia lamblia in a patient living with HIV and AIDS-a case report. *Scholars Journal of Applied Medical Sciences*, 3 (5C) (2015), 1975–1977.
9. L.-G. Tian, T.-P. Wang, S. Lv, F.-F. Wang, J. Guo, X.-M. Yin, Y.-C. Cai, M. K. Dickey, P. Steinmann, and J.-X. Chen. HIV and intestinal parasite co-infections among a Chinese population: An immunological profile. *Infectious Diseases of Poverty*, 2 (2013), 18.
10. S. Samko, A. Kilbas, and O. Marichev. *Fractional Integrals and Derivatives: Theory and Applications*. London, UK: Gordon and Breach Science Publishers, 1993.
11. M. Alquran, K. Al-Khaled, M. Ali, and O. A. Arqub. Bifurcations of the time-fractional generalized coupled Hirota-Satsuma KdV system. *Waves Wavelets and Fractals*, 3 (2017), 31–39.
12. D. Copot, R. De Keyser, E. Derom, M. Ortigueira, and C. M. Ionescu. Reducing bias in fractional order impedance estimation for lung function evaluation. *Biomedical Signal Processing and Control*, 39 (2018), 74–80.
13. R. M. Lizzy, K. Balachandran, and J. J. Trujillo. Controllability of nonlinear stochastic fractional neutral systems with multiple time varying delays in control. *Chaos, Solitons & Fractals*, 102 (2017), 162–167.
14. I. Podlubny. *Fractional Differential Equations: An Introduction to Fractional Derivatives, Fractional Differential Equations, to Methods of Their Solution and Some of Their Applications*. Amsterdam, the Netherlands: Elsevier, 1999.
15. S. G. Samko, A. A. Kilbas, and O. I. Marichev. *Fractional Integrals and Derivatives: Theory and Applications*. Yverdon, Switzerland: Gordon and Breach, 1993.
16. M. Caputo and M. Fabrizio. A new definition of fractional derivative without singular kernel. *Progress in Fractional Differentiation and Applications*, 1(2) (2015), 1–13.
17. S. Ullah, M. A. Khan, and M. Farooq. A new fractional model for the dynamics of the hepatitis B virus using the Caputo-Fabrizio derivative. *European Physical Journal Plus*, 133 (2018), 237. doi:10.1140/epjp/i2018-12072-4.
18. M. M. Khader and K. M. Saad. A numerical study by using the Chebyshev collocation method for a problem of biological invasion: Fractional Fisher equation. *International Journal of Biomathematics*, 11 (8) (2018), 1850099.
19. V. F. Morales-Delgado, et al. Analytic solution for oxygen diffusion from capillary to tissues involving external force effects: A fractional calculus approach. *Physica A: Statistical Mechanics and Its Applications*, 523 (2019), 48–65.
20. V. F. Morales-Delgado, et al. Application of the Caputo-Fabrizio and Atangana-Baleanu fractional derivatives to mathematical model of cancer chemotherapy effect. *Mathematical Methods in the Applied Sciences*, 42 (4) (2019), 1167–1193.

21. K. M. Saad, S. Deniz, and D. Baleanu. On a new modified fractional analysis of Nagumo equation. *International Journal of Biomathematics*, 12 (3) (2019), 1950034.
22. F. Gómez, J. Rosales, and M. Guía. RLC electrical circuit of non-integer order. *Open Physics*, 11 (10) (2013), 1361–1365.
23. J. F. Gómez-Aguilar, R. Razo-Hernández, and D. Granados-Lieberman. A physical interpretation of fractional calculus in observables terms: Analysis of the fractional time constant and the transitory response. *Revista mexicana de física*, 60 (1) (2014), 32–38.
24. J. F. Gómez-Aguilar, et al. Atangana-Baleanu fractional derivative applied to electromagnetic waves in dielectric media. *Journal of Electromagnetic Waves and Applications*, 30 (15) (2016), 1937–1952.
25. J. F. Gómez-Aguilar, et al. Modeling and simulation of the fractional space-time diffusion equation. *Communications in Nonlinear Science and Numerical Simulation*, 30 (1–3) (2016), 115–127.
26. J. F. Gómez-Aguilar. Space–time fractional diffusion equation using a derivative with nonsingular and regular kernel. *Physica A: Statistical Mechanics and Its Applications*, 465 (2017), 562–572.
27. K. M. Saad and J. F. Gómez-Aguilar. Analysis of reaction–diffusion system via a new fractional derivative with non-singular kernel. *Physica A: Statistical Mechanics and Its Applications*, 509 (2018), 703–716.
28. H. Yépez-Martínez and J. F. Gómez-Aguilar. A new modified definition of Caputo–Fabrizio fractional-order derivative and their applications to the Multi Step Homotopy Analysis Method (MHAM). *Journal of Computational and Applied Mathematics*, 346 (2019), 247–260.
29. A. Coronel-Escamilla, et al. Synchronization of chaotic systems involving fractional operators of Liouville–Caputo type with variable-order. *Physica A: Statistical Mechanics and Its Applications*, 487 (2017), 1–21.
30. J. Singh, D. Kumar, and A. Kilichman. Numerical solutions of nonlinear fractional partial differential equations arising in spatial diffusion of biological populations. *Abstract and Applied Analysis*, 2014 (2014), 12 pages.
31. K. M. Owolabi and A. Atangana. Analysis and application of new fractional Adams–Bashforth scheme with Caputo-Fabrizio derivative. *Chaos, Solitons & Fractals*, 105 (2017), 111–119.
32. M. Abdulhameed, D. Vieru, and R. Roslan. Magnetohydrodynamic electroosmotic flow of Maxwell fluids with Caputo-Fabrizio derivatives through circular tubes. *Computers & Mathematics with Applications*, 74 (2017), 2503–2519.
33. M. A. Firoozjaee, H. Jafari, A. Lia, and D. Baleanu. Numerical approach of FokkerPlanck equation with Caputo-Fabrizio fractional derivative using Ritz approximation. *Journal of Computational Applied Mathematics*, 339 (2018), 367–373.
34. K. A. AbroEmail, A. Ahmed, and M. A. Uqaili. A comparative mathematical analysis of RL and RC electrical circuits via Atangana-Baleanu and Caputo-Fabrizio fractional derivatives. *European Physical Journal Plus*, 133 (2018), 113.
35. M. A. Dokuyucu, E. Celik, H. Bulut, and H. M. Baskonus. Cancer treatment model with the Caputo-Fabrizio fractional derivative. *European Physical Journal Plus*, 133 (2018), 92.

36. J. Singh, D. Kumar, Z. Hammouch, and A. Atangana. A fractional epidemiological model for computer viruses pertaining to a new fractional derivative. *Applied Mathematics and Computation*, 316 (2018), 504–515.
37. A. Atangana and D. Baleanu. New fractional derivatives with non-local and nonsingular kernel: Theory and application to heat transfer model. *Thermal Science*, 89 (2016), 763–769.
38. A. Atangana and I. Koca. Chaos in a simple nonlinear system with Atangana-Baleanu derivatives with fractional order. *Chaos Solitons & Fractals*, 89 (2016), 447–454.
39. B. S. T. Alkahtani. Chuas circuit model with Atangana-Baleanu derivative with fractional order. *Chaos Solitons & Fractals*, 89 (2016), 547–551.
40. S. Ullah, M. Altaf Khan, and M. Farooq. Modeling and analysis of the fractional HBV model with Atangana-Baleanu derivative. *European Physical Journal Plus*, 133 (2018), 313. doi:10.1140/epjp/i2018-12120-1.
41. I. Koca. Modelling the spread of ebola virus with atangana-baleanu fractional 271 operators. *The European Physical Journal Plus*, 133 (3) (2018), 100.
42. A. A. Tateishi, H. V. Ribeiro, and E. K. Lenzi. The role of fractional time-derivative operators on anomalous diffusion. *Frontiers in Physics*, 5 (2017), 1–9.
43. A. Atangana. Non validity of index law in fractional calculus: A fractional differential operator with Markovian and non-Markovian properties. *Frontiers in Physics*, 505 (2017), 688–706.
44. A. Atangana and J. F. Gmez-Aguilar. Decolonisation of fractional calculus rules: Breaking commutativity and associativity to capture more natural phenomena. *European Physical Journal Plus*, 133 (2018), 166.
45. A. Atangana. On the new fractional derivative and application to nonlinear Fishers reaction-diffusion equation. *Applied Mathematics and Computation*, 273 (2016), 948–956.
46. M. Caputo and M. Fabrizio. On the notion of fractional derivative and applications to the hysteresis phenomena. *Meccanica* (2018). doi:10.1007/s11012-017-0652-y.
47. G. Certad, A. Arenas-Pinto, L. Pocaterra, G. Ferrara, J. Castro, A. Bello, and L. Nez. Cryptosporidiosis in HIV-infected Venezuelan adults is strongly associated with acute or chronic diarrhea. *American Journal of Tropical Medicine and Hygiene*, 73 (1) (2005), 54–57.
48. A. S. Kumurya and M. Y. Gwarzo. Cryptosporidiosis in HIV infected patients with diarrhoea in Kano state, North-western Nigeria. *Journal of AIDS and HIV Research*, 5 (8) (2013), 301–305.
49. C. N. Nkenfou, C. T. Nana, and V. K. Payne. Intestinal parasitic infections in HIV infected and non-infected patients in a low HIV prevalence region, West-Cameroon. *PLoS One*, 8 (2) (2013), e57914. doi:10.1371/journal.pone.0057914.
50. L.-G. Tian, J.-X. Chen, T.-P. Wang, et al. Co-infection of HIV and intestinal parasites in rural area of China. *Parasites and Vectors*, 5 (2012), 36.
51. P. Paboriboune, N. Phoumindr, E. Borel, et al. Intestinal parasitic infections in HIV-infected patients, Lao People's Democratic Republic. *PLoS One*, 9 (3) (2014), e91452. doi:10.1371/journal.pone.0091452.
52. K. O. Okosun, M. A. Khan, E. Bonyah, et al. On the dynamics of HIV-AIDS and cryptosporidiosis. *European Physical Journal Plus*, 132 (2017), 363. doi:10.1140/epjp/i2017-11625-3.

53. A. Atangana and D. Baleanu. New fractional derivatives with nonlocal and non-singular kernel: Theory and application to heat transfer model. arXiv preprint arXiv:1602.03408 (2016).
54. J. Losada and J. J. Nieto, Properties of the new fractional derivative without singular Kernel. *Progress in Fractional Differentiation and Applications*, 1 (2015), 87–92.
55. A. Atangana and K. M. Owolabi. New numerical approach for fractional differential equations. *Mathematical Modelling of Natural Phenomena*, 13 (1) (2018), 3.

8

A Fractional Mathematical Model to Study the Effect of Buffer and Endoplasmic Reticulum on Cytosolic Calcium Concentration in Nerve Cells

Brajesh Kumar Jha and Hardik Joshi

CONTENTS

8.1 Introduction 211
8.2 Preliminary of Fractional Calculus 213
8.3 Description of the Mathematical Model 214
    8.3.1 Calcium Buffering 214
    8.3.2 Endoplasmic Reticulum 216
8.4 Some Results of the Mathematical Model 217
8.5 Numerical Results and Discussion 221
    8.5.1 Spatial Variation of Ca^{2+} Distribution 221
    8.5.2 Temporal Variation of Ca^{2+} Distribution 222
    8.5.3 Variation of Ca^{2+} Distribution for Case 1 and Case 2 223
8.6 Conclusion 224
References 225

8.1 Introduction

Mathematical modelling and simulation in neuroscience have gained interest in researchers and scientists. It deals with the complex problem of intracellular and intracellular calcium (Ca^{2+}) signalling in nerve cells like neurons, astrocytes, etc. The advancement of software like MATLAB®, Mathematica, etc. encouraged the role of mathematics, to investigate the biophysical and physiological problems like synaptic transmission, dynamic regulation of cerebral microcirculation, neuronal activation, etc. [1,2].

The neuron is the basic building block of the nervous system. Many physiological processes take place inside the neurons, such as Ca^{2+} buffering, the flow of Ca^{2+} ions through a cell and Ca^{2+} flow through endoplasmic reticulum (ER). The Ca^{2+} signalling takes place in two ways: electrical signalling and chemical signalling [2].

The Ca^{2+} ion is one of the most important factors for the transmission of neurotransmitters. Ca^{2+} buffering modulates the intracellular calcium concentration ($[Ca^{2+}]$) level and plays a pivotal role in the Ca^{2+} signalling process. It reduces the free Ca^{2+} level and keeps the cell alive. Parkinson's disease (PD) is a prolonged brain disorder of the nervous system. The basic cause of the disease is the loss or dysfunction of dopamine neurons in the human brain. The symptoms of the disease are basically split into motor and non-motor symptoms. The motor symptoms are characterised by tremors, bradykinesia, and postural instability. The non-motor symptoms are anxiety, depression, loss of memory, constipation, etc [3].

In a literature survey, it is observed that many authors have studied Ca^{2+} distribution in different cells, such as neurons, astrocytes, fibrocytes, oocytes, myocytes, etc. [4–13]. Naik and Pardasani have studied Ca^{2+} distribution in the presence of voltage-gated calcium channel (VGCC), ryanodine receptors and buffers [6,7]. Pathak and Adlakha have developed a one-dimensional finite element model to study the physiological process of Ca^{2+} signalling in myocytes with the inclusion of external parameters like the excess buffer, pump, and leak [9]. They accomplished computationally that buffers play an important role in lowering down the $[Ca^{2+}]$, whereas the leak helps in rising up the same. Tewari and Pardasani have shown the diffusion of Ca^{2+} in the presence of excess buffer by using a finite element model [10]. Jha and Adlakha have used triangular elements to discretise the domain of dendritic spines [4,12]. They have used the finite element method to show the noteworthy effect of exogenous buffers, diffusion coefficients and influx on $[Ca^{2+}]$. Jha et al. have constructed a two-dimensional steady-state model using the finite element method, in which they have studied the effects of ethylene glycol-bis(β-aminoethyl ether)-N,N,N′,N′-tetraacetic acid (EGTA) and 1,2-bis(o-aminophenoxy)ethane-N,N,N′,N′-tetraacetic acid (BAPTA) buffers as well as the influx of $[Ca^{2+}]$ in astrocytes [13]. They have concluded that buffers like EGTA and BAPTA have a significant effect on Ca^{2+} distribution in astrocytes. Recently, a few attempts have been made to study the role of Ca^{2+} distribution on neurological disorders [14–17]. Also, it is reported in the literature that the ER is involved in the signalling process and works as a source of internal energy to protect the cells against homeostasis [18–22]. All the papers discussed the significant effect of various parameters like buffer and ER on Ca^{2+} distribution in nerve cells. It is observed that most of the research has been done using ordinary or partial differential equations. Hence, a very small amount of work is done to study calcium distribution in nerve cells with fractional operators. So in view of these, we developed a one-dimensional fractional reaction-diffusion model to study the effect of these various parameters on cytosolic $[Ca^{2+}]$ in nerve cells. Also, the obtained significant effects of these parameters are discussed in view of neurological disorders like PD.

In upcoming sections, some preliminary fractional calculus, the formation of a mathematical model, solution, results, and discussion are shown.

8.2 Preliminary of Fractional Calculus

Fractional calculus is an emerging branch of mathematics that deals with non-integer order derivative and integrals. Due to its amazing behaviour, it is preferred to modelling the biological model, epidemiological model, and their applications in science and engineering [23–30]. There are a variety of fractional derivatives that exist to deal with fractional-order differential equations. But among all, the Caputo derivative is considered here due to its well-understood physical behaviour. Some basic definitions are discussed below, which are useful to formulating the mathematical model and to finding the solution of the model [31–33].

Definition 8.1

For $\alpha > 0$ and $f : R^+ \to R$ the Riemann-Liouville integral is defined by

$$I^\alpha f(x) = \frac{1}{\Gamma(\alpha)} \int_0^x (x-t)^{\alpha-1} f(t) dt. \tag{8.1}$$

Definition 8.2

Let f be an absolutely continuous on $[0, t]$ and $\alpha \in (n-1, n)$, then the Riemann-Liouville derivative is defined by

$${}_0^{RL}D_t^\alpha f(t) = \frac{1}{\Gamma(n-\alpha)} \left(\frac{d}{dt}\right)^n \int_0^t (t-\tau)^{n-1-\alpha} f(\tau) d\tau. \tag{8.2}$$

Definition 8.3

For $n \in \mathbb{N}$ and $t > 0$, the Caputo derivative for a function f is defined by

$${}_0^C D_t^\alpha f(t) = \frac{1}{\Gamma(n-\alpha)} \int_0^t (t-\tau)^{n-1-\alpha} f^{(n)}(\tau) d\tau. \tag{8.3}$$

Definition 8.4

The Laplace transform for Caputo time derivative for a function f of order $\alpha \in (n-1, n)$ is defined by

$$L\left\{{}_0^C D_t^\alpha f(t); s\right\} = s^\alpha F(s) - \sum_{k=0}^{n-1} s^{\alpha-k-1} f^{(k)}(0). \tag{8.4}$$

Definition 8.5

The Fourier transform for Caputo space derivative for a function f of order $\beta \in (n-1, n)$ is defined by

$$F\{{}_{-\infty}^{C}D_x^\beta f(x); w\} = (-iw)^\beta F(w). \tag{8.5}$$

8.3 Description of the Mathematical Model

The theoretical analysis and formulation of the mathematical model of Ca^{2+} signalling is based on the fundamental conservation law in the differential form [34] (Figure 8.1).

The law states that the rate of change of mass within a domain is described as mass entering into the domain and mass exiting the domain. Besides this, due to chemical species inside the domain, we also look for a generation of mass and conservation of mass. Mathematically this phenomenon is expressed as

$$\frac{\partial M}{dt} - \frac{\partial J}{dx} = f(M, x, t) \tag{8.6}$$

where M represents the concentration of mass, J is the rate at which mass move from one end to another end at time t, and f is the net rate of mass per unit volume at location x and time t with the inclusion of production and consumption of mass. The generation of Ca^{2+} and the exchange of Ca^{2+} through the ER are considered in the proposed model.

8.3.1 Calcium Buffering

There are a number of proteins (buffers) found in neuron cells. Buffers stabilise the free Ca^{2+} ion concentration within the neuron cell. As the free Ca^{2+}

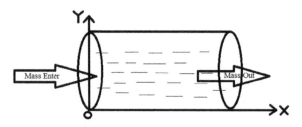

FIGURE 8.1
Mass transfer in domain.

A Fractional Mathematical Model

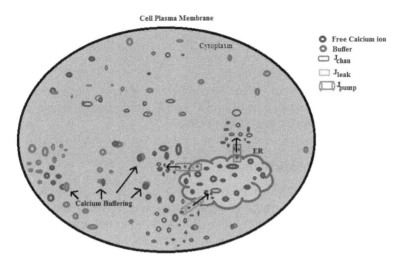

FIGURE 8.2
Ca²⁺ buffering process.

ions enter into the cell, they react with intracellular buffers and make Ca²⁺-bound buffers. In the dissociation process, it leaves free Ca²⁺ ions into the cell. Thus, buffers maintain the effect of the cytoplasm [Ca²⁺] by binding and releasing Ca²⁺ ions [35,36]. The reaction of Ca²⁺ and buffers are defined as below (Figure 8.2).

$$[Ca^{2+}] + [B] \underset{k^-}{\overset{k^+}{\Leftrightarrow}} [CaB]. \tag{8.7}$$

where [Ca²⁺], [B], and [CaB] are intracellular [Ca²⁺], free and bound buffers, respectively.

$$\frac{\partial [Ca^{2+}]}{\partial t} = D_{Ca} \frac{\partial^2}{\partial x^2}[Ca^{2+}] + \sum_j R_j, \tag{8.8}$$

$$\frac{\partial [B]}{\partial t} = D_B \frac{\partial^2}{\partial x^2}[B] + R_j, \tag{8.9}$$

$$\frac{\partial [CaB]}{\partial t} = D_{CaB} \frac{\partial^2}{\partial x^2}[CaB] - R_j, \tag{8.10}$$

where

$$R_j = -k^+ [B][Ca^{2+}] + k^- [CaB]. \qquad (8.11)$$

where D_{Ca}, D_B, and D_{CaB} are diffusion coefficients of free Ca^{2+}, free buffer, and Ca^{2+}-bound buffers, respectively, and k^+ and k^- are association and dissociation rate constants for buffer 'j', respectively.

8.3.2 Endoplasmic Reticulum

The ER is found inside the cell body of neuron cells. There are two types of ER found in the cell. One is rough endoplasmic reticulum (RER) and the other is smooth endoplasmic reticulum (SER). RER is found in neuron cells. The ER is known as the Ca^{2+} store, which actively controls the specific entry of free Ca^{2+} inside the cell. The model incorporates the gating kinetics of the ER channel as given by the Keizer and De Young [37]. Othmer and Tang simplified the Keizer-De Young model, which exhibits both excitability and frequency encoding [38]. The mathematical expression of J_{Leak}, J_{Chan}, and J_{Pump} is shown below [2,39].

$$J_{Leak} = D_{Leak} \left([Ca^{2+}]_{ER} - [Ca^{2+}]\right)$$

$$= \frac{D_{Leak}}{C_1}(1+C_1)\left(\frac{C_0}{1+C_1} - [Ca^{2+}]\right). \qquad (8.12)$$

where D_{Leak} is the diffusivity constant, C_1 is the volume ratio of ER and the cytosol, and $C_0 = C_1[Ca^{2+}]_{ER} + [Ca^{2+}]$.

$$J_{Chan} = P \cdot D_{Chan} \left([Ca^{2+}]_{ER} - [Ca^{2+}]\right)$$

$$= \frac{D_{Chan}}{C_1}(1+C_1)\left(\frac{C_0}{1+C_1} - [Ca^{2+}]\right). \qquad (8.13)$$

where D_{Chan} is the channel conductance, and p is the open channel probability.

$$J_{Pump} = P_R^{max} \frac{[Ca^{2+}]^2}{[Ca^{2+}]^2 + (K_R^M)^2}. \qquad (8.14)$$

where J_{Pump} is the inward flux modelled by a Michaelis-Menten expression with a Hill coefficient with n = 2 [2,39].

A Fractional Mathematical Model

Then by combining equations (8.1) through (8.14), the proposed mathematical model can be framed as

$${}_0^C D_t^\alpha [Ca^{2+}](x,t)$$
$$= D_{ca} \cdot {}_{-\infty}^C D_x^\beta [Ca^{2+}](x,t) - k_j^+[B]_\infty ([Ca^{2+}]-[Ca^{2+}]_\infty)(x,t) + J_{Leak} + J_{Chan} - J_{Pump},$$
(8.15)

where ${}_0^C D_t^\alpha [Ca^{2+}]$ and ${}_{-\infty}^C D_x^\beta [Ca^{2+}]$ are the Caputo time-fractional derivative of order α ($0 < \alpha \le 1$) and Caputo space-fractional derivative of order β ($1 < \beta \le 2$), respectively [31–33].

By inserting the Ca^{2+} fluxes into Equation (8.15), the governing equation for Ca^{2+} distribution in the nerve cells is

$$\begin{aligned}{}_0^C D_t^\alpha [Ca^{2+}] &= D_{ca} \cdot {}_{-\infty}^C D_x^\beta [Ca^{2+}] - k_j^+[B]_\infty ([Ca^{2+}]-[Ca^{2+}]_\infty) \\ &+ \frac{D_{Leak}}{C_1}(1+C_1)\left(\frac{C_0}{1+C_1}-[Ca^{2+}]\right) \\ &+ \frac{D_{Chan}}{C_1}(1+C_1)\left(\frac{C_0}{1+C_1}-[Ca^{2+}]\right) - P_R^{max}\frac{[Ca^{2+}]^2}{[Ca^{2+}]^2+(K_R^M)^2}.\end{aligned}$$
(8.16)

According to physiology of the cells, the initial and boundary conditions are

$$[Ca^{2+}](x,0) = g(x), \qquad [Ca^{2+}](\pm\infty,t) = 0.$$
(8.17)

where $g(x)$ is the sufficiently smooth and well-behaved function described as the Dirac delta function.

8.4 Some Results of the Mathematical Model

In this section, we obtain the results of the proposed model. To derive the main results, we need the following lemma:

Lemma 8.1

Let $C=[Ca^{2+}]$, $K=K_R^M$, $a_1=k_j^+[B]_\infty + \left((1+C_1)(D_{Leak}+P\cdot D_{Chan}) + \frac{P_R^{max}}{K}\right)$, $b_1 = k_j^+[B]_\infty C_\infty +$ $C_0(D_{Leak}+P\cdot D_{Chan})$ and $K \gg C$ then the model (8.16 and 8.17) is recast as

$${}_0^C D_t^\alpha C = D_{ca} \cdot {}_{-\infty}^C D_x^\beta C - a_1 C + b_1, \qquad C(x,0) = g(x), \qquad C(\pm\infty,t) = 0.$$
(8.18)

Proof

By considering the preceding fact, the model is written as

$$ {}_0^C D_t^\alpha C = D_{ca} \cdot {}_{-\infty}^C D_x^\beta C - k_j^+ [B]_\infty (C - C_\infty) + \frac{D_{Leak}}{C_1}(1+C_1)\left(\frac{C_0}{1+C_1} - C\right) $$
$$ + \frac{D_{Chan}}{C_1}(1+C_1)\left(\frac{C_0}{1+C_1} - C\right) - P_R^{max} \frac{C^2}{C^2+K^2}. \tag{8.19} $$

For $K \gg C$ which gives $\frac{C^2}{C^2+K^2} < \frac{C^2}{K^2} < \frac{C}{K}$. By incorporating these results, the model is reduced as

$$ {}_0^C D_t^\alpha C = $$

$$ D_{ca} \cdot {}_{-\infty}^C D_x^\beta C - k_j^+[B]_\infty(C-C_\infty) + (1+C_1)(D_{Leak} + P \cdot D_{Chan})\left(\frac{C_0}{1+C_1} - C\right) - P_R^{max}\frac{C}{K}. \tag{8.20} $$

Then for the sake of simplicity, Equation (8.20) is written as

$$ {}_0^C D_t^\alpha C = D_{ca} \cdot {}_{-\infty}^C D_x^\beta C - a_1 C + b_1. \tag{8.21} $$

Hence completing the proof.

Lemma 8.2

Let $C = [Ca^{2+}]$, $K = K_R^M$, $a_2 = k_j^+[B]_\infty + (1+C_1)(D_{Leak} + P \cdot D_{Chan})$, $b_2 = k_j^+[B]_\infty C_\infty + \left(C_0(D_{Leak} + P \cdot D_{Chan}) - \frac{P_R^{max}}{q^2+1}\right)$ and $K \ll C$ then the model in equations (8.16) and (8.17) is recast as

$$ {}_0^C D_t^\alpha C = D_{ca} \cdot {}_{-\infty}^C D_x^\beta C - a_2 C + b_2, \quad C(x,0) = g(x), \quad C(\pm\infty, t) = 0. \tag{8.22} $$

Proof

By considering the preceding fact, the model is written as

$$ {}_0^C D_t^\alpha C = D_{ca} \cdot {}_{-\infty}^C D_x^\beta C - k_j^+[B]_\infty(C-C_\infty) + \frac{D_{Leak}}{C_1}(1+C_1)\left(\frac{C_0}{1+C_1} - C\right) $$
$$ + \frac{D_{Chan}}{C_1}(1+C_1)\left(\frac{C_0}{1+C_1} - C\right) - P_R^{max}\frac{C^2}{C^2+K^2}. \tag{8.23} $$

A Fractional Mathematical Model

For $K \ll C$, take $K = qC$ where $0 < q < 1$ which gives $\frac{C^2}{C^2+K^2} = \frac{1}{q^2+1}$. By incorporating these results, the model is reduced as

$$_0^C D_t^\alpha C = D_{ca} \cdot {}_{-\infty}^C D_x^\beta C - k_j^+[B]_\infty(C - C_\infty) + (1+C_1)(D_{\text{Leak}} + P \cdot D_{\text{Chan}})$$
$$\left(\frac{C_0}{1+C_1} - C\right) - \frac{P_R^{\max}}{q^2+1}. \tag{8.24}$$

Then for sake of simplicity, Equation (8.24) is written as

$$_0^C D_t^\alpha C = D_{ca} \cdot {}_{-\infty}^C D_x^\beta C - a_2 C + b_2. \tag{8.25}$$

Hence completing the proof.
Now, we prove the main theorem.

Theorem 8.1

For $0 < \alpha \leq 1$, $1 < \beta \leq 2$, $a = a_i$, $b = b_i$, $\forall i = 1, 2$ and $\xi = \left\{D_{Ca}(-ik)^\beta - a\right\} \cdot t^\alpha$ then equations (8.21) and (8.25) are written as

$$_0^C D_t^\alpha C = D_{ca} \cdot {}_{-\infty}^C D_x^\beta C - aC + b, \quad C(x,0) = g(x), \quad C(\pm\infty, t) = 0. \tag{8.26}$$

which possess a unique solution in terms of the Mittag-Leffler function as

$$G_{\alpha,\beta}(x,t) = \frac{1}{2\pi}\int_{-\infty}^{\infty} e^{-ikx} E_\alpha(\xi)\, dk + \frac{bt^\alpha}{2\pi}\int_{-\infty}^{\infty} e^{-ikx} E_{\alpha,\alpha+1}(\xi)\, dk. \tag{8.27}$$

Proof

Applying the fractional temporal Laplace and spatial Fourier transform on Equation (8.26), we get

$$\hat{\tilde{C}}_{\alpha,\beta}(k,s) = \frac{s^{\alpha-1}}{s^\alpha - \left(D_{Ca}(-ik)^\beta - a\right)} \cdot \hat{g}(k) + \frac{b\delta(k)s^{-1}}{s^\alpha - \left(D_{Ca}(-ik)^\beta - a\right)}, \tag{8.28}$$

where $\hat{\tilde{C}}_{\alpha,\beta}$ is the fractional Laplace and Fourier transform of C. Then by inverting the fractional Laplace and Fourier transforms we get

$$G_{\alpha,\beta}(x,t) = \frac{1}{2\pi}\int_{-\infty}^{\infty} e^{-ikx}E_\alpha(\xi)dk + \frac{bt^\alpha}{2\pi}\int_{-\infty}^{\infty} e^{-ikx}E_{\alpha,\alpha+1}(\xi)dk, \qquad (8.29)$$

where $G(x,t)$ is Green's function, $E_\alpha(z) = \sum_{k=0}^{\infty}\frac{z^k}{\Gamma(\alpha k+1)}$, $R(\alpha) > 0$, $\alpha, z \in \mathbb{C}$, and $E_{\alpha,\beta}(z) = \sum_{k=0}^{\infty}\frac{z^k}{\Gamma(\alpha k+\beta)}$, $R(\alpha), R(\beta) > 0$, $\alpha, \beta, z \in \mathbb{C}$ are the Mittag-Leffler functions for one- and two-parameters, respectively [31–33].

Hence completing the proof.

The obtained solution is not in a closed form, so to get a closed-form solution further is split into time-fractional reaction-diffusion equations and space-fractional reaction-diffusion equations.

Corollary 8.1 (Temporal diffusion)

For $0 < \alpha \le 1$, $\beta = 2$ and $\psi = e^{-\left(\frac{x^2}{4D_{Ca}t^\alpha k} + at^\alpha k\right)}$ then the model in Equation (8.26) possess a well-posed solution in terms of a special case of the exponential function as

$$G_{\alpha,2}(x,t) = \frac{1}{2\sqrt{\pi D_{Ca}t^\alpha}}\int_0^\infty \psi \cdot k^{-\frac{1}{2}}M_\alpha(k)dk + \frac{bt^\alpha}{2\sqrt{\pi D_{Ca}t^\alpha}}\int_0^\infty \psi \cdot k^{-\frac{1}{2}}\phi(-\alpha,1;-k)dk, \quad (8.30)$$

where $M_\alpha(z) = \sum_{k=0}^{\infty}\frac{(-1)^k z^k}{\Gamma(-\alpha k+(1-\alpha))\cdot k!}$, $0 < \alpha < 1$ and $\phi(\alpha,\beta;z) = \sum_{k=0}^{\infty}\frac{z^k}{\Gamma(\alpha k+\beta)\bullet k!}$, $\alpha > -1$, $\beta \in \mathbb{C}$ are the Mainardi function and Wright function, respectively [31–33].

Proof

The proof is analogous with Theorem 8.1.

Corollary 8.2 (Spatial diffusion)

For $1 < \beta \le 2$, $\alpha = 1$ and $\varsigma = S\left(\frac{x}{(D_{Ca}t)^{1/\beta}}/\beta,1,1,0;1\right)$ then the model in Equation (8.26) possesses a well-posed solution in terms of stable distribution as

$$G_{1,\beta}(x,t) = \frac{e^{-at}}{(D_{Ca}t)^{1/\beta}}\varsigma + \frac{bt}{(D_{Ca}t)^{1/\beta}\{\ln\varsigma - at\}} \times \left[\left\{\frac{e^{-at}}{(D_{Ca}t)^{1/\beta}}\varsigma\right\} - 1\right]. \quad (8.31)$$

Proof

The proof is analogous with Theorem 8.1.

8.5 Numerical Results and Discussion

The values of physiological parameters used to simulate the results are given in Table 8.1 or stated along with the figure.

8.5.1 Spatial Variation of Ca^{2+} Distribution

Figure 8.3 shows the spatial variation of $[Ca^{2+}]$ in the presence of a buffer and ER. The value of the temporal order derivative in Figure 8.3 is $\alpha = 1$ and time $t = 0.01$ s. From Figure 8.3a and b, it is observed that increasing the amount of buffer concentration from 50 to 100 µM, decreases the profile of $[Ca^{2+}]$ as it moves farther away from the source. Physiologically, it happens due to a higher amount of buffer reacting with free Ca^{2+} ions in the cell and making the Ca^{2+}-bound buffer control the intracellular $[Ca^{2+}]$ environment. In Figure 8.3, $\alpha = 1$ and $\beta = 2$ are taken, and it means that the classical reaction-diffusion model validates our results with the previously known results [11,13]. The main aim of this study is to obtain the results at any fraction of time, which can be easily visualised in Figure 8.3. The fractional derivative operator portrayed the accurate results to help us understand the elevation in the intracellular $[Ca^{2+}]$.

Figure 8.4 shows the spatial variation of $[Ca^{2+}]$ in the presence of buffer and ER. The value of the spatial order derivative is $\beta = 2$ and $t = 0.01$ s. From Figure 8.4a and b, it is observed that a higher amount of buffer significantly decreases the profile of $[Ca^{2+}]$. Also, it is observed that the low order of time derivative attains a higher profile of $[Ca^{2+}]$. To increase the order of the time

TABLE 8.1

Values of Physiological Parameters

Symbol	Parameter	Value
D_{Ca}	Diffusion coefficient	250 µm²/s
$[B]$	Buffer concentration	50–100 µM
$[Ca^{2+}]_\infty$	Background $[Ca^{2+}]$	0.1 µM
k^+ EGTA	Buffer association rate	1.5 µM⁻¹s⁻¹
$[Ca^{2+}]_{ER}$	ER $[Ca^{2+}]$	500 µM
C_1	V_{ER}/V_{cyt}	0.185
P_R^{max}	Maximum Ca^{2+} uptake	0.9
D_{Leak}	Ca^{2+} leak flux	0.11 s⁻¹
D_{Chan}	Channel conductance	6 s⁻¹
K_R^M	Dissociation constant of Ca^{2+} pump	0.1 µM

Source: Keener, J. and Sneyd, J., *Mathematical Physiology*, 2nd ed., Interdisciplinary Applied Mathematics, Springer, New York, 2009; Jha, B.K. et al., *J. Comput.*, 3, 74–80, 2011.

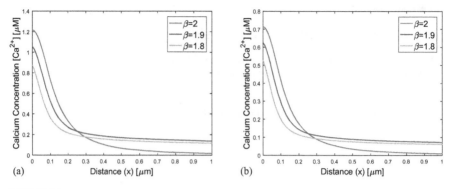

FIGURE 8.3
Spatial variation of [Ca^{2+}] in the presence of buffer and ER with $\alpha = 1$: (a) [B] = 50 µM and (b) [B] = 100 µM.

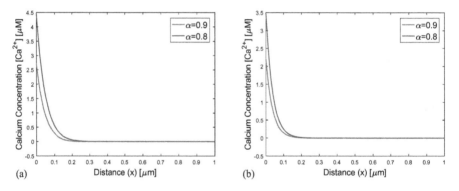

FIGURE 8.4
Spatial variation of [Ca^{2+}] in the presence of buffer and ER with $\beta = 2$: (a) [B] = 50 µM and (b) [B] = 100 µM.

derivative from fractional to ordinary derivative, the spatial variation profile of [Ca^{2+}] attains the behaviour of the classical diffusion equation, which can be easily shown in Figures 8.3 and 8.4. Thus, the fractional-order gives the exact flow profile of Ca^{2+} distribution in the nerve cells which reveals the elevation process of intracellular [Ca^{2+}].

8.5.2 Temporal Variation of Ca^{2+} Distribution

Figure 8.5 shows the temporal variation of [Ca^{2+}] in the presence of buffer and ER. The value of the temporal order derivative in Figure 8.5 is $\alpha = 1$ and $x = 0.001$ µm. From Figure 8.5, it is observed when $\alpha = 1$ and $\beta = 2$ (standard reaction-diffusion) the peak in the profile of [Ca^{2+}] is 1.39 and 0.88 µM, respectively, in Figure 8.5a and b. This decrease in the peak value in the

A Fractional Mathematical Model

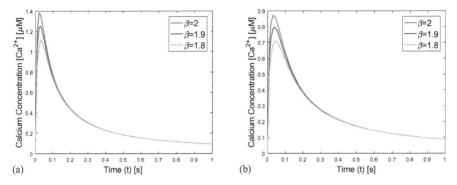

FIGURE 8.5
Temporal variation of [Ca²⁺] in the presence of buffer and ER with $\alpha = 1$: (a) [B] = 50 μM and (b) [B] = 100 μM.

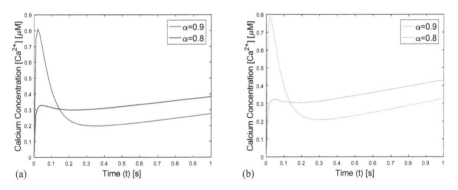

FIGURE 8.6
Temporal variation of [Ca²⁺] in the presence of buffer and ER with $\beta = 2$: (a) [B] = 50 μM and (b) [B] = 100 μM.

profile of [Ca²⁺] is due to the increased amount of buffer concentration, that is, 50 to 100 μM. Also, it is observed that as a decrease in the order of space derivative, the peak in the temporal variation of [Ca²⁺] is also decreased.

Figure 8.6 shows the temporal variation of [Ca²⁺] in the presence of buffer and ER. The value of the space order derivative in Figure 8.6 is $\beta = 2$ and $x = 0.001$ μm. From Figure 8.6, it is observed that the profile corresponding to the low-order time derivative attained a steady-state early as compared to high-order time derivative. This distinguishes the behaviour in the temporal variation of [Ca²⁺] as seen with the help of fractional calculus.

8.5.3 Variation of Ca²⁺ Distribution for Case 1 and Case 2

Figure 8.7 shows the spatial variation of [Ca²⁺] in the presence of buffer and ER. The value of the temporal and spatial order derivative is $\alpha = 0.9$ and $\beta = 2$,

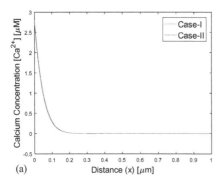 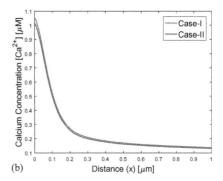

FIGURE 8.7
Spatial variation of [Ca²⁺] in the presence of buffer and ER with [B] = 50 µM: (a) $\alpha = 0.9$ and $\beta = 2$ (b) $\alpha = 1$ and $\beta = 1.9$.

and $\alpha = 1$ and $\beta = 1.9$, respectively, in Figure 8.7a and b. The amount of buffer concentration is taken as 50 µM to visualise the effect of two cases [case-1 $K \gg C$ (Lemma 8.1)] and [case-2 $K \ll C$ (Lemma 8.2)] considered in this study to handle the non-linearity occurring in the diffusion model. The time span is $t = 0.01$ s. Both Figures 8.7a and b give the spatial variation of [Ca²⁺] corresponding to different cases. It is observed that the Ca²⁺ distribution profile corresponding to both the cases is in positive correlation and almost provides the same physiological impact on nerve cells to protect them from toxicity.

8.6 Conclusion

The present study reveals the complex interplay occurring between the buffer and ER in the form of a fractional diffusion model. The fractional diffusion model is considered rather than a classical diffusion model due to its non-local behaviour. The fractional derivative provides the subtle changes that occur in the Ca²⁺ distribution profile of nerve cells. The solution is obtained for two cases [case-1 $K \gg C$ (Lemma 8.1)] and [case-2 $K \ll C$ (Lemma 8.2)], to smoothly deal with the non-linearity that occurred in the diffusion model due to ER flux. The results are obtained corresponding to the time-fractional and space-fractional diffusion model. It is observed that the buffer and ER flux provide a significant effect on [Ca²⁺]. ER flux maintains the adequate intracellular [Ca²⁺] in the nerve cells. An alteration in this complex interplay leads to the early symptoms of chronic PD. By incorporating more physiological parameters like voltage-gated Ca²⁺ channels and sodium Ca²⁺ exchangers, the present model can be extended to study the more complex behaviour of the nerve cells.

References

1. J. D. Murray, *Mathematical Biology II – Spatial Models and Biomedical Applications*, 3rd ed. Interdisciplinary Applied Mathematics, Springer, New York, 2008.
2. J. Keener and J. Sneyd, *Mathematical Physiology*, 2nd ed. Interdisciplinary Applied Mathematics, Springer, New York, 2009.
3. M. Brini, T. Calì, D. Ottolini, and E. Carafoli, "Neuronal calcium signaling: Function and dysfunction," *Cell. Mol. Life Sci.*, vol. 71, no. 15, pp. 2787–2814, 2014.
4. A. Jha and N. Adlakha, "Analytical solution of two dimensional unsteady state problem of calcium diffusion in a neuron cell," *J. Med. Imaging Heal. Inform.*, vol. 4, no. 4, pp. 547–553, 2014.
5. M. Kotwani, "Modeling and simulation of calcium dynamics in fibroblast cell involving excess buffer approximation (EBA), ER flux and SERCA pump," *Procedia Comput. Sci.*, vol. 49, pp. 347–355, 2015.
6. P. A. Naik and K. R. Pardasani, "One dimensional finite element model to study calcium distribution in oocytes in presence of VGCC, RyR and buffers," *J. Med. Imaging Heal. Inform.*, vol. 5, pp. 471–476, 2015.
7. P. A. Naik and K. R. Pardasani, "Finite element model to study calcium distribution in oocytes involving voltage gated Ca^{2+} channel, ryanodine receptor and buffers," *Alexandria J. Med.*, vol. 52, March, pp. 43–49, 2016.
8. S. Panday and K. R. Pardasani, "Finite element model to study the mechanics of calcium regulation in oocyte," *J. Mech. Med. Biol.*, vol. 14, no. 2, pp. 1–16, 2014.
9. K. Pathak and N. Adlakha, "Finite element model to study two dimensional unsteady state calcium distribution in cardiac myocytes," *Alexandria J. Med.*, vol. 52, no. 3, pp. 261–268, 2016.
10. S. G. Tewari and K. R. Pardasani, "Finite element model to study two dimensional unsteady state cytosolic calcium diffusion in presence of excess buffers," *IAENG Int. J. Appl. Math.*, vol. 40, no. 3, pp. 1–5, 2010.
11. B. K. Jha, N. Adlakha, and M. N. Mehta, "Finite volume model to study the effect of buffer on cytosolic Ca^{2+} advection diffusion," *Int. J. Math. Comput. Phys. Electr. Comput. Eng.*, vol. 5, no. 3, pp. 248–251, 2011.
12. A. Jha and N. Adlakha, "Finite element model to study the effect of exogenous buffer on calcium dynamics in dendritic spines," *Int. J. Model. Simulation, Sci. Comput.*, vol. 5, no. 2, pp. 1–12, 2014.
13. B. K. Jha, N. Adlakha, and M. N. Mehta, "Two-dimensional finite element model to study calcium distribution in astrocytes in presence of excess buffer," *Int. J. Biomath.*, vol. 7, no. 3, pp. 1–11, 2014.
14. D. D. Dave and B. K. Jha, "Delineation of calcium diffusion in alzheimeric brain," *J. Mech. Med. Biol.*, vol. 18, no. 2, pp. 1–15, 2018.
15. B. K. Jha, H. Joshi, and D. D. Dave, "Portraying the effect of calcium-binding proteins on cytosolic calcium concentration distribution fractionally in nerve cells," *Interdiscip. Sci.*, vol. 10, no. 4, pp. 674–685, 2018.
16. H. Joshi and B. K. Jha, "Fractional reaction diffusion model for Parkinson's disease," in: D. Pandian, X. Fernando, Z. Baig, F. Shi (eds) *Proceedings of the International Conference on ISMAC in Computational Vision and Bio-Engineering (ISMAC-CVB)*. ISMAC 2018. Lecture Notes in Computational Vision and Biomechanics, vol. 30, Springer, Cham, pp. 1739–1748, 2018.

17. H. Joshi and B. K. Jha, "Fractionally delineate the neuroprotective function of calbindin-28k in Parkinson's disease," *Int. J. Biomath.*, vol. 11, no. 8, p. 1850103, 2018.
18. E. J. Ryu, H. P. Harding, J. M. Angelastro, O. V. Vitolo, D. Ron, and L. A. Greene, "Endoplasmic reticulum stress and the unfolded protein response in cellular models of Parkinson's disease," *J. Neurosci.*, vol. 22, no. 24, pp. 10690–10698, 2002.
19. T. Omura, M. Kaneko, Y. Okuma, K. Matsubara, and Y. Nomura, "Endoplasmic reticulum stress and Parkinson's disease: The role of HRD1 in averting apoptosis in neurodegenerative disease," *Oxid. Med. Cell. Longev.*, vol. 2013, pp. 1–7, 2013.
20. C. Hetz, G. Mercado, and P. Valde, "An ERcentric view of Parkinson's disease," *Cell*, vol. 19, no. 3, pp. 165–175, 2013.
21. M. Puzianowska-kuznicka and J. Kuznicki, "The ER and ageing II: Calcium homeostasis," *Ageing Res. Rev.*, vol. 8, pp. 160–172, 2009.
22. S. Tsujii, M. Ishisaka, H. Hara, M. Ishisaka, and H. Hara, "Modulation of endoplasmic reticulum stress in Parkinson's disease," *Eur. J. Pharmacol.*, vol. 765, pp. 154–156, 2015.
23. B. Yaşkıran and M. Yavuz, "Approximate-analytical solutions of cable equation using conformable fractional operator," *New Trends Math. Sci.*, vol. 4, no. 5, pp. 209–219, 2017.
24. M. Yavuz and B. Yaşkıran, "Conformable derivative operator in modelling neuronal dynamics," *Appl. Appl. Math.*, vol. 13, no. 2, pp. 803–817, 2018.
25. J. Singh, D. Kumar, Z. Hammouch, and A. Atangana, "A fractional epidemiological model for computer viruses pertaining to a new fractional derivative," *Appl. Math. Comput.*, vol. 316, pp. 504–515, 2018.
26. M. Yavuz and N. Özdemir, "Comparing the new fractional derivative operators involving exponential and Mittag-Leffler kernel," *Discret. Contin. Dyn. Syst.*, vol. 13, no. 3, pp. 995–1006, 2018.
27. M. Yavuz and E. Bonyah, "New approaches to the fractional dynamics of schistosomiasis disease model," *Physica A*, vol. 525, pp. 373–393, 2019.
28. D. Kumar, J. Singh, S. D. Purohit, and R. Swroop, "A hybrid analytical algorithm for nonlinear fractional wave-like equations," *Math. Model. Nat. Phenom.*, vol. 14, no. 3, p. 304, 2019.
29. D. Kumar, J. Singh, A. Prakash, and R. Swroop, "Numerical simulation for system of time-fractional linear and nonlinear differential equations," *Prog. Fract. Differ. Appl.*, vol. 5, no. 1, pp. 65–77, 2019.
30. A. Goswami, J. Singh, D. Kumar, and S. Rathore, "An efficient analytical approach for fractional equal width equations describing hydro-magnetic waves in cold plasma," *Physica A*, vol. 524, pp. 563–575, 2019.
31. I. Podlubny, *Fractional Differential Equations*. Academic Press, New York, 1999.
32. K. Diethelm, *The Analysis of Fractional Differential Equations: An Application-Oriented Exposition Using Differential Operators of Caputo Type*. Springer-Verlag, Berlin, Germany, 2010.
33. D. Baleanu, K. Diethelm, E. Scalas, and J. J. Trujillo, Fractional Calculus: Models and numerical methods, Series on Complexity, Nonlinearity and Chaos, vol. 3, World Scientific, New Jersey, 2012.
34. J. Crank, *The Mathematics of Diffusion*, 2nd ed. Oxford University Press, London, UK, 1975.
35. J. Gliabert, "Cytoplasmic calcium buffering," *Adv. Exp. Med. Biol.*, vol. 740, no. 3, p. e304, 2012.

36. B. Schwaller, "Cytosolic Ca^{2+} buffers," *Cold Spring Harb. Perspect. Biol.*, vol. 2, pp. 1–20, 2010.
37. J. Keizer and G. W. De Young, "Two roles for Ca^{2+} in agonist stimulated Ca^{2+} oscillations," *Biophys. J.*, vol. 61, no. 3, pp. 649–660, 1992.
38. H. G. Othmer and Y. Tang, "Oscillations and waves in a model of InsP3-controlled calcium dynamics," in P.K. Maini, J.D. Murray (eds.), *Experimental and Theoretical Advances in Biological Pattern Formation*, NATO ASI Series (Series A: Life Sciences), vol. 259, Springer, Boston, MA, pp. 277–300, 1993.
39. B. K. Jha, N. Adlakha, and M. N. Mehta, "Finite volume model to study the effect of ER on cytosolic calcium distribution in astrocytes," *J. Comput.*, vol. 3, no. December, pp. 74–80, 2011.

9

Fractional SIR Epidemic Model of Childhood Disease with Mittag-Leffler Memory

P. Veeresha, D. G. Prakasha and Devendra Kumar

CONTENTS

9.1 Introduction 229
9.2 Preliminaries 232
9.3 Mathematical Model of the Childhood Disease in AB
    Fractional Derivatives 234
9.4 Existence of Solutions for the Fractional Model 235
9.5 Numerical Approximations 240
9.6 Numerical Simulations 242
9.7 Conclusion 244
References 245

9.1 Introduction

An epidemic is the swift evolution of infectious disease in a given population to a huge number of people within a small period of time. In the eighteenth century, Swiss physicist and mathematician Bernoulli proposed and cultivated the concept of mathematical modelling for the evolution of disease [1], which provides the origin to the development of modern epidemiology. Later in the twentieth century, Ross et al. [2] established the modelling of infectious disease and elucidated the nature of epidemic models by the help of the law of mass action. Recently, epidemic models are widely applied to analyse the epidemiological processes that contain the contagious disease transmission.

It is basic and essential for health proprietors to protect adolescents from disease that can be prevented by inoculation. Even though anticipatory vaccines reduce the prevalence of infectious illnesses among the children, epidemic diseases in the adolescents are the imperative public health problem. In order to resolve and overcome the defects, we frequently model these diseases with the help of mathematics. The mathematical model can aid us in the transmission and dynamical behaviour of childhood diseases [3–8].

Recently, the childhood diseases have become the most deliberate infective diseases. Rubella, poliomyelitis, and measles are the more familiar childhood diseases [9,10]. These diseases usually influence children because the population of children is extremely large, and they are prone to these diseases more so than adults [9]. Particularly, measles is an extremely virulent disease that is instigated by a respiratory infection of the Morbilli virus. In addition, the population can be mainly distributed into two classes: mature and premature populations. The premature population considers the fixed duration to become mature, which is called a maturation delay. In the dynamics of disease, the diseases don't spread rapidly but rather they require some duration in the body, which known as the latent period of the disease. However, after the World Health Organization (WHO) started the Expanded Program on Immunization [11], the effort of vaccination was widely started to all children beginning in the year 1974. In this connection, the mathematical models play a vital role in understanding the nature of transmission of the diseases and helps us to analyse the behaviour of disease-affected children. Further, the mathematical models can help us to capture the growth of the diseases, and these models can produce diverse methods to control its propagation.

The kernel of mathematical tools for demonstrating the practical difficulties exist in real life is as old as the conception of the world. The development of science and technology is magnetizing the considerations of the authors with the help of mathematical models to understand, describe, and predict the future behavior of the natural phenomena. A mathematical model is a representation of a system with the aid of mathematical theories, rules, formulas, and methods. Humankind has invented the most influential mathematical concepts known as calculus with the integral and differential operators, which can model and simulate numerous mechanisms that have arisen in environments of past decades. Recently, many researchers pointed out that classical derivatives fail to capture essential physical properties like long-range, anomalous diffusion, random walk, non-Markovian processes, and most importantly heterogeneous behaviours. Hence, many scientists and mathematicians find out that the classical differential operators are not always suitable tools to model the non-linear phenomena.

The concept of fractional calculus originated in 1695 as a question of extension meaning, but recently it attracted the attention of many young researchers due to its ability to describe the various simulating and interesting properties related to time and history consequences. The most intriguing leaps in scientific and engineering applications have been found within the framework of fractional calculus (FC). From the last few decades, it becomes the most powerful instrument in examining and describing the non-linear complex biological model, particularly the model that describes the dynamics of human diseases. This is due to favourable properties such as analyticity, hereditary, non-locality, and memory effect. The concept of derivative with arbitrary order has been coined due to the complexities connected to a phenomenon with heterogeneities. The differential operators with non-integer order are capable of capturing essential behaviour of the complex media having diffusion mechanisms. Due to the rapid growth of computer and mathematical techniques and software, many authors started working on the arbitrary order calculus with fundamentals and its applications.

The basic notion of fractional order derivatives and integrals are prescribed by many pioneers in order to understand the behaviour of natural processes that arise in daily life [12–17]. FC has been associated with practical ventures, and it has been expansively applied to human diseases, chaos theory, optics, nanotechnology, and other areas [18–26]. For instance, authors in [27] considered the fractional derivative in order to capture the behaviour of the imaginary part of the obtained solution for Zakharov equations with distinct fractional order, the model of Casson fluid based on generalised Fourier's and Fick's laws analysed by authors in [28] within the framework of fractional calculus, and they presented some interesting consequences. Researchers in [29] consider the fractional order biochemical reaction model and find its approximate analytical solution. They also showed the importance of fractional calculus—the behaviour of the approximated analytical solution for the model describing the propagation of shallow water waves captured in terms of plots with a coupled surface for different fractional orders and some interesting consequences, which help to understand the nature of shallow water waves with the influence of time and history properties.

Recently, mathematicians Atangana and Baleanu (AB) introduced and nurtured a new form of derivative and integrals with fractional order. These concepts are defined with the aid of extended Mittag-Leffler function as the non-singular and non-local kernel. Further AB derivatives accept all the essential properties of fractional derivatives. Recently, many scientists have used these derivatives to analyse diverse classes of models in comparison with the classical order as well as other fractional derivatives. They showed that the AB derivative is more effective while analysing the nature and physical behaviour of the problems [30–49]. Previously defined fractional operators were unable to satisfy the conditions required to fulfil all the essential and important properties. For instance, the Riemann-Liouville operator fails to present the physical meaning of the initial condition, and it is unable to satisfy the classical condition of the derivative of constant function. Later, Caputo derived a new operator that satisfies the preceding cited limitations, and it has been widely used as a fractional operator in order to study various physical and biological models. Recently, many mathematicians and physicists pointed out some drawbacks while examining some specific properties of the model arising in thermal science and fluid mechanics. In order to overcome these obliges, Caputo and Fabrizio (CF) derived new fractional operator. But recently, Atangana and some other authors pointed out the limitation of the CF operator related to no-singular and non-local kernels. These are the essential properties that play vital roles while exemplifying some specific and essential physical behaviour of the phenomena. In the present scenario, the AB derivate has attracted many researchers. For instance, authors in [50] considered newly introduced derivatives in order to capture the behaviour of the exothermic reactions model having a constant heat source in porous media with power exponential and Mittag-Leffler laws, and they presented some simulating results. The numerical solution of the predator-prey model with the Mittag-Leffler kernel has been investigated by authors in [51], and many authors have considered a newly defined novel fractional operator in order to examine a wild class of complex and non-linear systems describing various interesting phenomena related to science and technology.

The biological models that are modelled and described with the help of arbitrary order differential equations have been demonstrated in analysing the behaviour of susceptible-infected-recovered (SIR) epidemic models. These models are developed and analysed by many researchers using many techniques with the aid of classical and fractional order derivatives [52–54]. Moreover, many authors have established the arbitrary order models to analyse the diffusion equation to describe the effect of these diseases. In the present framework, we analyse the proposed childhood disease model using the AB derivative and find the solution with the Adams-Bashforth scheme. Further, the considered model is described with the help of a new fractional derivative incorporated in the memory effect, the non-singular as well as non-local kernel, and these properties are very essential while analysing the behaviour of human diseases.

In the present investigation, we applied a novel technique in order to find the solution for the fractional SIR epidemic model of childhood disease and presented the behaviour of the obtained solution for two different cases within the framework of Mittag-Leffler memory. The rest of the chapter is arranged as: In Section 9.2, the fundamentals of fractional order integrals, derivatives, and some important results are presented and which will be used in the in the present investigation. Section 9.3 is concerned with mathematical model of the SIR epidemic model of childhood disease in the AB fractional derivative. In Section 9.4, existence of solutions for the future model is presented. In Section 9.5, the solution for the considered model presented in the system (9.11) is demonstrated. In Sections 9.6 and 9.7, the numerical simulation and concluding remarks are demonstrated, respectively.

9.2 Preliminaries

In this segment, we present the fundamental notion of AB derivatives and integrals, which are essential in the present framework [55].

Definition 9.1

The Atangana-Baleanu-Caputo (ABC) fractional derivative for a function $f \in H^1(a,b) (b > a,\ \mu \in [0,1))$ is presented as follows:

$$_a^{ABC}D_t^\mu (f(t)) = \frac{B(\mu)}{1-\mu} \int_a^t f'(\vartheta) E_\mu \left[\mu \frac{(t-\vartheta)^\mu}{\mu-1}\right] d\vartheta. \tag{9.1}$$

Definition 9.2

The AB derivative of fractional order for a function $f \in H^1(a,b), b > a,\ \mu \in [0,1]$ in the Riemann-Liouville sense is presented as follows:

$$_a^{ABR}D_t^\mu (f(t)) = \frac{B(\mu)}{1-\mu} \frac{d}{dt} \int_a^t f(\vartheta) E_\mu \left[\mu \frac{(t-\vartheta)^\mu}{\mu-1}\right] d\vartheta. \tag{9.2}$$

Definition 9.3

The fractional AB integral related to the non-local kernel is presented as

$$_a^{AB}I_t^\mu (f(t)) = \frac{1-\mu}{B(\mu)} f(t) + \frac{\mu}{B(\mu)\Gamma(\mu)} \int_a^t f(\vartheta)(t-\vartheta)^{\mu-1} d\vartheta. \tag{9.3}$$

Theorem 9.1

The following conditions hold true for a continuous function f defined on $[a,b]$ [55]:

$$\left\| _a^{ABC}D_t^\mu f(t) \right\| < \frac{B(\mu)}{1-\mu} \|f(x)\|, \text{ where } \|f(x)\| = \max_{0 \leq x \leq 1} |f(x)|. \tag{9.4}$$

Theorem 9.2

The following Lipschitz conditions, respectively, hold true for both Riemann-Liouville and AB derivatives defined in Equations (9.1) and (9.2) [55],

$$\left\| _a^{ABC}D_t^\mu f_1(t) - _a^{ABC}D_t^\mu f_2(t) \right\| < K_1 \|f_1(x) - f_2(x)\|, \tag{9.5}$$

and

$$\left\| _a^{ABC}D_t^\mu f_1(t) - _a^{ABC}D_t^\mu f_2(t) \right\| < K_2 \|f_1(x) - f_2(x)\|. \tag{9.6}$$

Theorem 9.3

The time-fractional differential equation ${}^{ABC}_{a}D^{\mu}_{t}f_1(t) = s(t)$ has a unique solution and which is defined as [55]

$$f(t) = \frac{1-\mu}{\text{ABC}(\mu)}s(t) + \frac{\mu}{\text{ABC}(\mu)\Gamma(\mu)}\int_0^t s(\varsigma)(t-\varsigma)^{\mu-1}\,d\varsigma. \quad (9.7)$$

9.3 Mathematical Model of the Childhood Disease in AB Fractional Derivatives

In this segment, we consider the system of three fractional order differential equations describing the epidemic model of childhood disease. Let the total population size at time t be represented by $N(t)$. In the proposed model, a susceptible group, an infected group, and quarantined or removed group are respectively symbolised by $S(t)$, $I(t)$, and $R(t)$. This model also accepts that the effectiveness of the vaccine as 100%. Further, the natural death charge μ in the population is not equal to the birth rates; hence, the population size N is not accurately constant. Here, p ($0 < p < 1$) fraction of vaccinated population at birth of each years, and citizens are born into the population at a fixed birth rate α. We also consider the susceptible individual will transfer into the infected group via contact with an infected individual. An average contact rate is denoted by β, and the infected individual recovery rate is represented by δ. Now, we consider the system of differential equations elucidating the preceding phenomenon [56]

$$\frac{dS}{dt} = (1-p)\alpha - \beta S(t)I(t) - \alpha S(t), \quad (9.8)$$

$$\frac{dI}{dt} = \beta S(t)I(t) - (\delta + \alpha)I(t), \quad (9.9)$$

$$\frac{dR}{dt} = p\alpha + \delta I(t) - \alpha R(t). \quad (9.10)$$

In order to incorporate the effect of hereditary, non-locality, and memory effect, we modified the preceding system by introducing the AB fractional derivative in the place of the time derivative, and which are presented as follows:

$${}^{ABC}_{0}D^{\mu}_{t}[S(t)] = (1-p)\alpha - \beta S(t)I(t) - \alpha S(t),$$

$${}^{ABC}_{0}D^{\mu}_{t}\left[I(t)\right]=\beta S(t)I(t)-(\delta+\alpha)I(t), 0<\mu\leq 1,$$

$${}^{ABC}_{0}D^{\mu}_{t}\left[R(t)\right]=p\alpha+\delta I(t)-\alpha R(t),\qquad(9.11)$$

with initial conditions

$$S(0)=N_1,\ I(0)=N_2,\ R(0)=N_3. \qquad(9.12)$$

9.4 Existence of Solutions for the Fractional Model

Here, we used the fixed-point theorem in order to examine the existence of the solution for the suggested fractional order mathematical model. Since the considered model presented in the system in Equation (9.11) is non-local as well as complex, there are no particular algorithms or techniques that exist to derive the exact solutions. However, under some particular conditions, the existence of the solution is assured. Now, the system in Equation (9.11) is considered as follows:

$$\begin{cases} {}^{ABC}_{0}D^{\mu}_{t}\left[S(t)\right]=\mathcal{G}_1(t,S), \\ {}^{ABC}_{0}D^{\mu}_{t}\left[I(t)\right]=\mathcal{G}_2(t,I), \\ {}^{ABC}_{0}D^{\mu}_{t}\left[R(t)\right]=\mathcal{G}_3(t,R). \end{cases} \qquad(9.13)$$

The foregoing system is transformed to the Volterra integral equation with the aid of Theorem 9.3, and which is as follows:

$$\begin{cases} S(t)-S(0)=\dfrac{(1-\mu)}{ABC(\mu)}\mathcal{G}_1(t,S)+\dfrac{\mu}{ABC(\mu)\Gamma(\mu)}\int_0^t \mathcal{G}_1(\zeta,S)(t-\zeta)^{\mu-1}d\zeta, \\ I(t)-I(0)=\dfrac{(1-\mu)}{ABC(\mu)}\mathcal{G}_2(t,I)+\dfrac{\mu}{ABC(\mu)\Gamma(\mu)}\int_0^t \mathcal{G}_2(\zeta,I)(t-\zeta)^{\mu-1}d\zeta, \\ R(t)-R(0)=\dfrac{(1-\mu)}{ABC(\mu)}\mathcal{G}_3(t,R)+\dfrac{\mu}{ABC(\mu)\Gamma(\mu)}\int_0^t \mathcal{G}_3(\zeta,R)(t-\zeta)^{\mu-1}d\zeta. \end{cases} \qquad(9.14)$$

Theorem 9.4

The kernel $\mathcal{G}_1$ satisfies the Lipschitz condition and contraction if the condition $0\leq(\beta\lambda_1+\alpha)<1$ holds.

Proof

In order to prove the desired result, we assume the two functions S and S_1, then

$$\|\mathcal{G}_1(t,S) - \mathcal{G}_1(t,S_1)\| = \|-\{\beta I(t)\}[S(t) - S(t_1)] - \alpha[S(t) - S(t_1)]\|$$
$$\leq \{\beta \|I(t)\| + \alpha\} \|S(t) - S(t_1)\| \quad (9.15)$$
$$\leq \{\beta \lambda_1 + \alpha\} \|S(t) - S(t_1)\|.$$

Putting $\eta_1 = \beta \lambda_1 + \alpha$ (where $\|I(t)\| \leq \lambda_1$ be the bounded function) in the preceding inequality, then we have

$$\|\mathcal{G}_1(t,S) - \mathcal{G}_1(t,S_1)\| \leq \eta_1 \|S(t) - S(t_1)\|. \quad (9.16)$$

This gives the Lipschitz condition is obtained for $\mathcal{G}_1$. Further we can see that if $0 \leq (\beta \lambda_1 + \alpha) < 1$, then it implies the contraction. The remaining cases can be verified in the similar manner, and which are given as follows:

$$\|\mathcal{G}_2(t,I) - \mathcal{G}_2(t,I_1)\| \leq \eta_2 \|I(t) - I(t_1)\|,$$
$$\|\mathcal{G}_3(t,R) - \mathcal{G}_3(t,R_1)\| \leq \eta_3 \|R(t) - R(t_1)\|. \quad (9.17)$$

The recursive form of Equation (9.14) is defined as follows:

$$\begin{cases} S_n(t) = \dfrac{(1-\mu)}{\text{ABC}(\mu)} \mathcal{G}_1(t, S_{n-1}) + \dfrac{\mu}{\text{ABC}(\mu)\Gamma(\mu)} \int_0^t \mathcal{G}_1(\zeta, S_{n-1})(t-\zeta)^{\mu-1} d\zeta, \\ I_n(t) = \dfrac{(1-\mu)}{\text{ABC}(\mu)} \mathcal{G}_2(t, I_{n-1}) + \dfrac{\mu}{\text{ABC}(\mu)\Gamma(\mu)} \int_0^t \mathcal{G}_2(\zeta, I_{n-1})(t-\zeta)^{\mu-1} d\zeta, \quad (9.18) \\ R_n(t) = \dfrac{(1-\mu)}{\text{ABC}(\mu)} \mathcal{G}_3(t, R_{n-1}) + \dfrac{\mu}{\text{ABC}(\mu)\Gamma(\mu)} \int_0^t \mathcal{G}_3(\zeta, R_{n-1})(t-\zeta)^{\mu-1} d\zeta. \end{cases}$$

The associated initial conditions are

$$S_0(t) = S(0), \; I_0(t) = I(0), \; R_0(t) = R(0).$$

Fractional SIR Epidemic Model of Childhood Disease with Mittag-Leffler 237

The successive difference between the terms is presented as follows:

$$\begin{cases} \phi_{1n}(t) = S_n(t) - S_{n-1}(t) \\ = \dfrac{(1-\mu)}{ABC(\mu)} \big(\mathcal{G}_1(t, S_{n-1}) - \mathcal{G}_1(t, S_{n-2})\big) + \dfrac{\mu}{ABC(\mu)\Gamma(\mu)} \int_0^t \mathcal{G}_1(\zeta, S_{n-1})(t-\zeta)^{\mu-1} d\zeta, \\ \phi_{2n}(t) = I_n(t) - I_{n-1}(t) \\ = \dfrac{(1-\mu)}{ABC(\mu)} \big(\mathcal{G}_2(t, I_{n-1}) - \mathcal{G}_2(t, I_{n-2})\big) + \dfrac{\mu}{ABC(\mu)\Gamma(\mu)} \int_0^t \mathcal{G}_2(\zeta, I_{n-1})(t-\zeta)^{\mu-1} d\zeta, \\ \phi_{3n}(t) = R_n(t) - R_{n-1}(t) \\ R_n(t) = \dfrac{(1-\mu)}{ABC(\mu)} \big(\mathcal{G}_3(t, R_{n-1}) - \mathcal{G}_3(t, R_{n-2})\big) \\ \qquad + \dfrac{\mu}{ABC(\mu)\Gamma(\mu)} \int_0^t \mathcal{G}_3(\zeta, R_{n-1})(t-\zeta)^{\mu-1} d\zeta. \end{cases} \quad (9.19)$$

Notice that

$$\begin{cases} S_n(t) = \sum_{i=1}^{n} \phi_{1i}(t), \\ I_n(t) = \sum_{i=1}^{n} \phi_{2i}(t), \\ R_n(t) = \sum_{i=1}^{n} \phi_{3i}(t). \end{cases} \quad (9.20)$$

By using Equation (9.16), after applying the norm on the first equation of the system in Equation (9.19), one can get

$$\|\phi_{1n}(t)\| \leq \dfrac{(1-\mu)}{ABC(\mu)} \eta_1 \|\phi_{1(n-1)}(t)\| + \dfrac{\mu}{ABC(\mu)\Gamma(\mu)} \eta_1 \int_0^t \|\phi_{1(n-1)}(\zeta)\| d\zeta. \quad (9.21)$$

Similarly, we have

$$\begin{cases} \|\phi_{2n}(t)\| \leq \dfrac{(1-\mu)}{ABC(\mu)} \eta_2 \|\phi_{2(n-1)}(t)\| + \dfrac{\mu}{ABC(\mu)\Gamma(\mu)} \eta_2 \int_0^t \|\phi_{2(n-1)}(\zeta)\| d\zeta, \\ \|\phi_{3n}(t)\| \leq \dfrac{(1-\mu)}{ABC(\mu)} \eta_3 \|\phi_{3(n-1)}(t)\| + \dfrac{\mu}{ABC(\mu)\Gamma(\mu)} \eta_3 \int_0^t \|\phi_{3(n-1)}(\zeta)\| d\zeta. \end{cases} \quad (9.22)$$

We prove the following theorem by making use of the preceding results.

Theorem 9.5

The solution for the system in Equation (9.11) will exist and is unique, if we have specific t_0 then

$$\dfrac{(1-\mu)}{ABC(\mu)} \eta_i + \dfrac{\mu}{ABC(\mu)\Gamma(\mu)} \eta_i < 1,$$

for $i = 1, 2, 3$.

Proof

Let us consider the bounded functions $S(t)$, $I(t)$, and $R(t)$ satisfying the Lipschitz condition. Then, by Equations (9.21) and (9.22), we have

$$\|\phi_{1i}(t)\| \leq \|S_n(0)\| \left[\dfrac{(1-\mu)}{ABC(\mu)} \eta_1 + \dfrac{\mu}{ABC(\mu)\Gamma(\mu)} \eta_1 \right]^n,$$

$$\|\phi_{2i}(t)\| \leq \|I_n(0)\| \left[\dfrac{(1-\mu)}{ABC(\mu)} \eta_2 + \dfrac{\mu}{ABC(\mu)\Gamma(\mu)} \eta_2 \right]^n, \quad (9.23)$$

$$\|\phi_{3i}(t)\| \leq \|R_n(0)\| \left[\dfrac{(1-\mu)}{ABC(\mu)} \eta_3 + \dfrac{\mu}{ABC(\mu)\Gamma(\mu)} \eta_3 \right]^n.$$

Therefore, the continuity as well as existence for the obtained solutions is proved. Subsequently, in order to show that the system in Equation (9.23) is the solution for the system in Equation (9.11), we consider

$$S(t) - S(0) = S_n(t) - B_{1n}(t),$$
$$I(t) - I(0) = I_n(t) - B_{2n}(t), \quad (9.24)$$
$$R(t) - R(0) = R_n(t) - B_{3n}(t).$$

Fractional SIR Epidemic Model of Childhood Disease with Mittag-Leffler

In order to obtain require result, we consider

$$\|B_{1n}(t)\| = \left\| \frac{(1-\mu)}{ABC(\mu)} \big(\mathcal{G}_1(S,t) - \mathcal{G}_1(S_{n-1},t)\big) \right.$$

$$\left. + \frac{\mu}{ABC(\mu)\Gamma(\mu)} \int_0^t (t-\zeta)^{\mu-1} \big(\mathcal{G}_1(\zeta,S) - \mathcal{G}_1(\zeta,S_{n-1})\big) d\zeta \right\|$$

$$\leq \frac{(1-\mu)}{ABC(\mu)} \|\big(\mathcal{G}_1(S,t) - \mathcal{G}_1(S_{n-1},t)\big)\| + \frac{\mu}{ABC(\mu)\Gamma(\mu)} \int_0^t \|\mathcal{G}_1(\zeta,S) - \mathcal{G}_1(\zeta,S_{n-1})\| d\zeta$$

$$\leq \frac{(1-\mu)}{ABC(\mu)} \eta_1 \|S - S_{n-1}\| + \frac{\mu}{ABC(\mu)\Gamma(\mu)} \eta_1 \|S - S_{n-1}\| t. \tag{9.25}$$

Similarly, at t_0 we can obtain

$$\|B_{1n}(t)\| \leq \left(\frac{(1-\mu)}{ABC(\mu)} + \frac{\mu t_0}{ABC(\mu)\Gamma(\mu)} \right)^{n+1} \eta_1^{n+1} M. \tag{9.26}$$

As n approaches to ∞, we can see that form in Equation (9.26), $\|B_{1n}(t)\|$ tends to 0. Similarly, we can verify for $\|B_{2n}(t)\|$ and $\|B_{3n}(t)\|$.

Next, it is necessity to demonstrate uniqueness for the solution of considered model. Suppose, $S^*(t)$, $I^*(t)$, and $R^*(t)$ be the set of other solutions, then we have

$$S(t) - S^*(t) = \frac{(1-\mu)}{ABC(\mu)} \big(\mathcal{G}_1(t,S) - \mathcal{G}_1(t,S^*)\big)$$

$$+ \frac{\mu}{ABC(\mu)\Gamma(\mu)} \int_0^t \big(\mathcal{G}_1(\zeta,S) - \mathcal{G}_1(\zeta,S^*)\big) d\zeta. \tag{9.27}$$

On applying norm, Equation (9.27) simplifies to

$$\|S(t) - S^*(t)\|$$

$$= \left\| \frac{(1-\mu)}{ABC(\mu)} \big(\mathcal{G}_1(t,S) - \mathcal{G}_1(t,S^*)\big) + \frac{\mu}{ABC(\mu)\Gamma(\mu)} \int_0^t \big(\mathcal{G}_1(\zeta,S) - \mathcal{G}_1(\zeta,S^*)\big) d\zeta \right\| \tag{9.28}$$

$$\leq \frac{(1-\mu)}{ABC(\mu)} \eta_1 \|S(t) - S^*(t)\| + \frac{\mu}{ABC(\mu)\Gamma(\mu)} \eta_1 t \|S(t) - S^*(t)\|.$$

On simplification

$$\|S(t)-S^*(t)\|\left(1-\frac{(1-\mu)}{ABC(\mu)}\eta_1 - \frac{\mu}{ABC(\mu)\Gamma(\mu)}\eta_1 t\right) \leq 0. \quad (9.29)$$

From preceding condition, it is clear that $S(t) = S^*(t)$, if

$$\left(1-\frac{(1-\mu)}{ABC(\mu)}\eta_1 - \frac{\mu}{ABC(\mu)\Gamma(\mu)}\eta_1 t\right) \geq 0. \quad (9.30)$$

Hence, Equation (9.30) evidences our essential result.

9.5 Numerical Approximations

The Adams-Bashforth has been recognised as a powerful technique and provides a highly accurate solution for the proposed models. Initially, this algorithm was introduced to solve integer order differential equations by applying the fundamental theorem of calculus, and then later, it extended to the concept of fractional differentiation for Riemann-Liouville (RL) and Caputo derivatives. In 2018, authors propose a novel Adams-Bashforth for new and distinct fractional derivatives [52]. The new technique considered the non-linearity of the kernels counting the power law for RL derivative, the exponential decay law for the CF derivative and Mittag-Leffler for AB derivative [57]. The stability and convergence of the obtained solution for three types of fractional derivatives has been effectively presented. Further, the fractional Fisher's equation is considered in order to present the efficiency of the future scheme, and authors effectively presented the numerical simulation for the considered methods in order to confirm the effect of each derivative.

In this segment, the solution for the proposed model cited in the system in Equation (9.11) is demonstrated. Here, we applied the Adams-type Bashforth scheme to find the solution for the considered SIR epidemic model. We consider the system in Equation (9.11) in the fractional Volterra form, in order to present the solution procedure of the future technique. By the assist of the fundamental theorem of integration, the first term of the system in Equation (9.11) becomes

$$S(t) - S(0) = \frac{1-\mu}{ABC(\mu)} \mathcal{G}_1(t,S) + \frac{\mu}{ABC(\mu)\Gamma(\mu)} \int_0^t \mathcal{G}_1(\zeta,S)(t-\zeta)^{\mu-1} d\zeta.$$

Fractional SIR Epidemic Model of Childhood Disease with Mittag-Leffler

For $t = t_{n+1}$, $n = 0,1,2,\ldots$ we have

$$S_{n+1}(t) - S(0) = \frac{1-\mu}{ABC(\mu)} \mathcal{G}_1(t_n, S_n) + \frac{\mu}{ABC(\mu)\Gamma(\mu)} \int_0^{t_{n+1}} \mathcal{G}_1(\zeta, S)(t_{n+1} - \zeta)^{\mu-1} d\zeta.$$

The successive difference between the terms is defined as follows:

$$S_{n+1}(t) - S_n = \frac{1-\mu}{ABC(\mu)} \left[\mathcal{G}_1(t_n, S_n) - \mathcal{G}_1(t_{n-1}, S_{n-1}) \right] + \frac{\mu}{ABC(\mu)\Gamma(\mu)} (K_{\mu,1}, K_{\mu,2}), \quad (9.31)$$

where

$$K_{\mu,1} = \int_0^{t_{n+1}} \mathcal{G}_1(t, S)(t_{n+1} - t)^{\mu-1} dt,$$

$$K_{\mu,2} = \int_0^{t_n} \mathcal{G}_1(t, S)(t_n - t)^{\mu-1} dt.$$

By applying the two-point interpolation polynomial, we approximate the $\mathcal{G}_1(t, S)$ as follows:

$$L(t) \simeq \mathcal{G}_1(t, S) = \frac{t - t_{n-1}}{\hbar} \mathcal{G}_1(t_n, x(t_n)) - \frac{t - t_n}{\hbar} \mathcal{G}_1(t_{n-1}, x(t_{n-1})), \quad (9.32)$$

where $\hbar = t_n - t_{n-1}$. Thus, by considering Equation (9.32) we get

$$K_{\mu,1} = \frac{\mathcal{G}_1(t_n, S_n)}{\hbar} \left[-\frac{2\hbar}{\mu} t_{n+1}^\mu + \frac{t_{n+1}^{\mu+1}}{\mu+1} \right] + \frac{\mathcal{G}_1(t_{n-1}, S_{n-1})}{\hbar} \left[\frac{\hbar}{\mu} t_{n+1}^\mu - \frac{t_{n+1}^{\mu+1}}{\mu+1} \right]. \quad (9.33)$$

Similarly,

$$K_{\mu,2} = \frac{\mathcal{G}_1(t_n, S_n)}{\hbar} \left[-\frac{\hbar}{\mu} t_n^\mu + \frac{t_n^{\mu+1}}{\mu+1} \right] + \frac{\mathcal{G}_1(t_{n-1}, S_{n-1})}{\hbar(\mu+1)} t_n^{\mu+1}. \quad (9.34)$$

By considering Equations (9.33) and (9.34) in Equation (9.31), we obtain

$$S_{n+1}(t) = S_n + \mathcal{G}_1(t_n, S_n) \left[\frac{1-\mu}{ABC(\mu)} \right.$$
$$+ \frac{\mu}{ABC(\mu)\hbar\,\Gamma(\mu)} \left(-\frac{2\hbar}{\mu} t_{n+1}^\mu + \frac{t_{n+1}^{\mu+1}}{\mu+1} + \frac{\hbar}{\mu} t_n^\mu - \frac{t_n^{\mu+1}}{\mu+1} \right) \right] \quad (9.35)$$
$$+ \mathcal{G}_1(t_{n-1}, S_{n-1}) \left[\frac{\mu-1}{ABC(\mu)} + \frac{\mu}{ABC(\mu)\hbar\,\Gamma(\mu)} \left(\frac{\hbar}{\mu} t_{n+1}^\mu - \frac{t_{n+1}^{\mu+1}}{\mu+1} - \frac{t_n^{\mu+1}}{\mu+1} \right) \right].$$

Similarly,

$$I_{n+1}(t) = I_n + \mathcal{G}_2(t_n, I_n)\left[\frac{1-\mu}{\text{ABC}(\mu)}\right.$$

$$+ \frac{\mu}{\text{ABC}(\mu)\hbar\Gamma(\mu)}\left(-\frac{2\hbar}{\mu}t_{n+1}^\mu + \frac{t_{n+1}^{\mu+1}}{\mu+1} + \frac{\hbar}{\mu}t_n^\mu - \frac{t_n^{\mu+1}}{\mu+1}\right)\right] \tag{9.36}$$

$$+ \mathcal{G}_2(t_{n-1}, I_{n-1})\left[\frac{\mu-1}{\text{ABC}(\mu)} + \frac{\mu}{\text{ABC}(\mu)\hbar\Gamma(\mu)}\left(\frac{\hbar}{\mu}t_{n+1}^\mu - \frac{t_{n+1}^{\mu+1}}{\mu+1} - \frac{t_n^{\mu+1}}{\mu+1}\right)\right],$$

$$R_{n+1}(t) = R_n + \mathcal{G}_3(t_n, R_n)\left[\frac{1-\mu}{\text{ABC}(\mu)}\right.$$

$$+ \frac{\mu}{\text{ABC}(\mu)\hbar\Gamma(\mu)}\left(-\frac{2\hbar}{\mu}t_{n+1}^\mu + \frac{t_{n+1}^{\mu+1}}{\mu+1} + \frac{\hbar}{\mu}t_n^\mu - \frac{t_n^{\mu+1}}{\mu+1}\right)\right] \tag{9.37}$$

$$+ \mathcal{G}_3(t_{n-1}, R_{n-1})\left[\frac{\mu-1}{\text{ABC}(\mu)} + \frac{\mu}{\text{ABC}(\mu)\hbar\Gamma(\mu)}\left(\frac{\hbar}{\mu}t_{n+1}^\mu - \frac{t_{n+1}^{\mu+1}}{\mu+1} - \frac{t_n^{\mu+1}}{\mu+1}\right)\right].$$

The convergence and stability analysis for the obtained solutions in Equations (9.35) through (9.37) can be directly conformed by the help of the same technique verified in Theorems 3.4 and 3.5 in [38].

9.6 Numerical Simulations

In order to illustrate that the future scheme is efficient and reliable, the solutions have been evaluated in this chapter for the considered childhood disease. In the present investigation, we considered distinct explanatory cases that are analysed by authors in [51] in order to analyse the behaviour of the considered SIR epidemic model with AB fractional derivative, and the parameters and initial value for the proposed model are defined in Table 9.1. The explorative nature of obtained solutions of the susceptible group $S(t)$, infected group $I(t)$, and removed or

TABLE 9.1
The Effect of the Vaccination for Various Parameters

Case	S_0	I_0	R_0	α	β	η	p	Comments
I	1	0	0	0.8	0.003	0.4	0.9	Disease eradication
II	0.8	0.2	0	0.8	0.03	0.4	0.0	No eradication

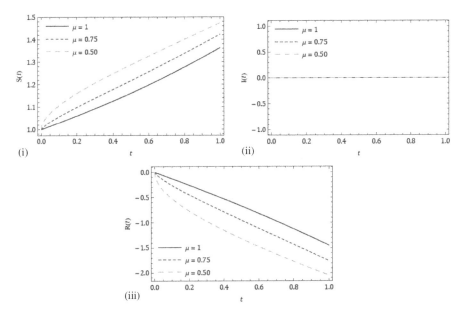

FIGURE 9.1
Response of obtained solution with respect to t for (i) $S(t)$, (ii) $I(t)$, and (iii) $R(t)$ at the value of the parameters defined in the Table 9.1 for **Case I** with diverse μ.

quarantined group $R(t)$ with a distinct Brownian motion and standard motion ($\mu = 1$) for Case I and Case II, respectively, are captured in the Figures 9.1 and 9.2. From cited plots, we can absorb that the considered fractional order epidemic model has a huge degree of flexibility. Moreover, the consequence of the proposed model evidently depends on its history and the parameters.

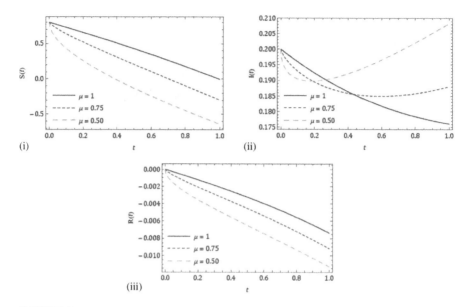

FIGURE 9.2
Behaviour of obtained solution with respect to t for (i) $S(t)$, (ii) $I(t)$, and (iii) $R(t)$ at the value of the parameters defined in the Table 9.1 for **Case II** with diverse μ.

9.7 Conclusion

In this chapter, we investigated the SIR model of childhood disease with the AB fractional derivative using the Adams-Bashforth scheme. Since, AB derivatives and integrals having fractional order, they are defined with the aid of the Mittag-Leffler function as the non-singular and non-local kernels, and the present investigation illuminates the effectiveness of the considered derivative. The dynamic nature of projected model depends on both the time history as well as the time instant, which can be proficiently modelled with the aid of the theory of fractional calculus. The graphical illustration conforms to the model and conspicuously depends on the considered parameters and the fractional order, which can stimulate the stability of the model. Finally, we conclude that the considered fractional operator is more effective as well as highly methodical. Hence, these derivatives can be applied to analyse the behaviour as well as predict the dynamical growth of human diseases.

References

1. Bernoulli, D. (1766). Essaid'une nouvelle analyse de la mortalitecausee par la petite verole. *Mem. Math. Phys. Acad. R. Sci.*, 1–41.
2. Ross, R., Annett, H. E. & Austen, E. E. (1900). Report of the malaria expedition of the Liverpool School of Tropical Medicine and Medical Parasitology. University Press of Liverpool, 1–71. doi:10.5962/bhl.title.99657.
3. Jansen, H. & Twizell, E. H. (2002). An unconditionally convergent discretization of the SEIR model. *Math. Comput. Simulation*, 58(2), 147–158.
4. Rohani, P., Earn, D. J., Finkenstadt, B. & Grenfell, B. T. (1998). Population dynamic interference among childhood diseases. *Proc. R. Soc. Lond.*, 265, 2033–2041.
5. Earn, D. J. D., Rohani, P., Bolker, B. M. & Grenfell, B. T. (2000). A simple model for complex dynamical transitions in epidemics. *Science*, 287(5453), 667–670.
6. Gumel, A. B., Moghadas, S. M., Yuan, Y. & Yu, P. (2003). Bifurcation and stability analyses of a 3-D SEIV model using normal form reduction and numerical simulation. *DCDIS*, 10(2), 317–330.
7. Anderson, R. M. & May, R. M. (1979). Population biology of infectious diseases: Part I. *Nature*, 280, 361–367.
8. May, R. M. & Anderson, R. M. (1979). Population biology of infectious diseases II. *Nature*, 280, 455–461.
9. Bai, M. & Ren, L. (2015). An SEIV epidemic model for childhood diseases with partial permanent immunity. *Comput. Math. Methods Med.*, 1–13. doi:10.1155/2015/420952.
10. Singh, H., Dhar, J., Bhatti, H. S. & Chandok, S. (2016). An epidemic model of childhood disease dynamics with maturation delay and latent period of infection. *Model. Earth Syst. Environ.*, 2(79). doi:10.1007/s40808-016-0131-9.
11. Henderson, R. H. (1984). The expanded programme on immunization of the world health organization. *Rev. Infect. Dis.*, 6(2), 475–479.
12. Liouville, J. (1832). Memoire surquelques questions de geometrieet de mecanique, etsur un nouveau genre de calcul pour resoudreces questions. *J. Ecole. Polytech.* 13, 1–69.
13. Riemann, G. F. B., Versucheinerallgemeinen Auffassung der Integration und Differentiation, Gesammelte Mathematische Werke, Leipzig, 1896.
14. Caputo, M., *Elasticita e Dissipazione*, Zanichelli, Bologna, 1969.
15. Miller, K. S. & Ross, B., *An Introduction to Fractional Calculus and Fractional Differential Equations*, Wiley, New York, 1993.
16. Podlubny, I., *Fractional Differential Equations*, Academic Press, New York, 1999.
17. Kilbas, A. A., Srivastava, H. M. & Trujillo, J. J., *Theory and Applications of Fractional Differential Equations*, Elsevier, Amsterdam, the Netherlands, 2006.
18. Baleanu, D., Guvenc, Z. B. & Tenreiro Machado, J. A., *New Trends in Nanotechnology and Fractional Calculus Applications*, Springer, New York, 2010.
19. Esen, A., Sulaiman, T. A., Bulut, H. & Baskonus, H. M. (2018). Optical solitons and other solutions to the conformable space-time fractional Fokas-Lenells equation. *Optik*, 167, 150–156.

20. Veeresha, P., Prakasha, D. G. & Baskonus, H. M. (2019). Solving smoking epidemic model of fractional order using a modified homotopy analysis transform method. *Math. Sci.*, 13(2), 115–128.
21. Baleanu, D., Wu, G. C. & Zeng, S. D. (2017). Chaos analysis and asymptotic stability of generalized Caputo fractional differential equations. *Chaos Soliton. Fract.*, 102, 99–105.
22. Veeresha, P., Prakasha, D. G. & Baskonus, H. M. (2019). New numerical surfaces to the mathematical model of cancer chemotherapy effect in Caputo fractional derivatives. *Chaos*, 29(013119). doi:10.1063/1.5074099.
23. Baskonus, H. M., Sulaiman, T. A. & Bulut, H. (2019). On the new wave behavior to the Klein-Gordon-Zakharov equations in plasma physics. *Indian J. Phys.*, 93(3), 393–399.
24. Veeresha, P., Prakasha, D. G. & Baskonus, H. M. (2019). Novel simulations to the time-fractional Fisher's equation. *Math. Sci.*, 13(1), 33–42.
25. Veeresha, P., Prakasha, D. G. & Baleanu, D. (2019). An efficient numerical technique for the nonlinear fractional Kolmogorov-Petrovskii-Piskunov equation. *Mathematics*, 7(3), 1–18. doi:10.3390/math7030265.
26. Singh, J. (2019). A new analysis for fractional rumor spreading dynamical model in a social network with Mittag-Leffler law. *Chaos*, 29(013137). doi:10.1063/1.5080691s.
27. Veeresha, P. & Prakasha, D. G. (2020). Solution for fractional generalized Zakharov equations with Mittag-Leffler function. *Results Eng.*, 5. doi:10.1016/j.rineng.2019.100085.
28. Sheikh, N. A., Ching, D. L. C., Khan, I., Kumar, D. & Nisar, K. S. (2019). A new model of fractional Casson fluid based on generalized Fick's and Fourier's laws together with heat and mass transfer. *Alexandria Eng. J.* doi:10.1016/j.aej.2019.12.023.
29. Veeresha, P., Prakasha, D. G. & Baskonus, H. M. (2020). An efficient technique for coupled fractional Whitham-Broer-Kaup equations describing the propagation of shallow water waves. *Adv. Intel. Syst. Comput.*, 49–75. doi:10.1007/978-3-030-39112-6_4.
30. Baskonus, H. M. (2019). Analysis of the dynamics of hepatitis E virus using the Atangana-Baleanu fractional derivative. *Eur. Phys. J. Plus*, 134(241), 1–11. doi:10.1140/epjp/i2019-12590-5.
31. Singh, J., Kumar, D., Hammouch, Z. & Atangana, A. (2018). A fractional epidemiological model for computer viruses pertaining to a new fractional derivative. *Appl. Math. Comput.*, 316, 504–515.
32. Atangana, A. & Alkahtani, B. T. (2015). Analysis of the Keller-Segel model with a fractional derivative without singular kernel. *Entropy*, 17, 4439–4453.
33. Atangana, A. & Alkahtani, B. T. (2016). Analysis of non-homogenous heat model with new trend of derivative with fractional order. *Chaos Soliton. Fract.*, 89, 566–571.
34. Singh, J., Kumar, D. & Kilichman, A. (2014). Numerical solutions of nonlinear fractional partial differential equations arising in spatial diffusion of biological populations. *Abstr. Appl. Anal.* Article ID 535793, 1–12. doi:10.1155/2014/535793.
35. Choudhary, A., Kumar, D. & Singh, J. (2014). Analytical solution of fractional differential equations arising in fluid mechanics by using Sumudu transform method. *Nonlinear Eng.*, 3(3), 133–139.
36. Uçar, S., Uçar, E., Özdemir, N. & Hammouch, Z. (2019). Mathematical analysis and numerical simulation for a smoking model with Atangana–Baleanu derivative. *Chaos Soliton. Fract.*, 118, 300–306.

37. Veeresha, P., Prakasha, D. G. & Kumar, D. (2020). An efficient technique for nonlinear time-fractional Klein-Fock-Gordon equation. *Appl. Math. Comput.*, 364. doi:10.1016/j.amc.2019.124637.
38. Jarad, F., Abdeljawad, T. & Hammouch, Z. (2018). On a class of ordinary differential equations in the frame of Atangana–Baleanu fractional derivative. *Chaos Soliton. Fract.*, 117, 16–20.
39. Owolabi, K. M. & Hammouch, Z. (2019). Mathematical modeling and analysis of two-variable system with noninteger-order derivative. *Chaos*, 29(013145). doi:10.1063/1.5086909.
40. Veeresha, P., Prakasha, D. G., Qurashi, M. A. & Baleanu, D. (2019). A reliable technique for fractional modified Boussinesq and approximate long wave equations. *Adv. Differ. Equ.*, 253, 1–23. doi:10.1186/s13662-019-2185-2.
41. Yokuş, A., Sulaiman, T. A., Gulluoglu, M. T. & Bulut, H. (2018). Stability analysis, numerical and exact solutions of the (1+1)-dimensional NDMBBM equation. *ITM Web of Conferences, EDP Sciences*, 22, 1–10. doi:10.1051/itmconf/20182201064.
42. Yokuş, A. (2018). Comparison of Caputo and conformable derivatives for time-fractional Korteweg–de Vries equation via the finite difference method. *Int. J. Mod. Phys. B*, 32(29). doi:10.1142/S0217979218503654.
43. Yokuş, A. & Tuz, M. (2017). An application of a new version of (G'/G)-expansion method. *AIP Conf. Proc.*, 020165. doi:10.1063/1.4972757.
44. Veeresha, P. & Prakasha, D. G. (2019). Solution for fractional Zakharov–Kuznetsov equations by using two reliable techniques. *Chin. J. Phys.*, 60, 313–330.
45. Sulaiman, T. A., Bulut, H., Yokus, A. & Baskonus, H. M. (2019). On the exact and numerical solutions to the coupled Boussinesq equation arising in ocean engineering. *Indian J. Phy.*, 93(5), 647–656.
46. Prakasha, D. G., Veeresha, P. & Rawashdeh, M. S. (2019). Numerical solution for (2+1)-dimensional time-fractional coupled Burger equations using fractional natural decomposition method. *Math. Meth. Appl. Sci.*, 42(10), 3409–3427.
47. Yang, A.-M., Zhang, Y. Z., Cattani, C., Xie, G. N., Rashidi, M. M., Zhou, Y. J. & Yang, X. J. (2014). Application of local fractional series expansion method to solve Klein-Gordon equations on cantor sets. *Abstr. Appl. Anal.*, 372741. doi:10.1155/2014/372741.
48. Prakash, A., Prakasha, D. G. & Veeresha, P. (2019). A reliable algorithm for time-fractional Navier-Stokes equations via Laplace transform. *Nonlinear Eng.*, 8, 695–701.
49. Heydari, M. H., Hooshmandasl, M. R., Mohammadi, F. & Cattani, C. (2014). Wavelets method for solving systems of nonlinear singular fractional Volterraintegro-differential equations. *Commun. Nonlinear Sci. Numer. Simul.*, 19(1), 37–48.
50. Kumar, D., Singh, J., Tanwar, K. & Baleanu, D. (2019). A new fractional exothermic reactions model having constant heat source in porous media with power, exponential and Mittag-Leffler laws. *Int. J. Heat Mass Tran*, 138, 1222–1227.
51. Ghanbari, B. & Kumar, D. (2019). Numerical solution of predator-prey model with Beddington-DeAngelis functional response and fractional derivatives with Mittag-Leffler kernel. *Chaos*, 29(6). doi:10.1063/1.5094546.
52. El-Shahed, M. & El-Naby, F. A. (2014). Fractional calculus model for childhood diseases and vaccines. *Appl. Math. Sci.*, 8(98), 4859–4866.

53. Selvam, A. G. M., Vianny, D. A. & Jacintha, M. (2018). Stability in a fractional order SIR epidemic model of childhood diseases with discretization. *J. Phys. Conf. Series*, 1139. doi:10.1088/1742-6596/1139/1/012009.
54. Arafa, A. A. M., Rida, S. Z. & Khalil, M. (2012). Solutions of fractional order model of childhood diseases with constant vaccination strategy. *Math. Sci. Lett.*, 1(1), 17–23.
55. Atangana, A. & Baleanu, D. (2016). New fractional derivatives with non-local and non-singular kernel, theory and application to heat transfer model. *Therm. Sci.*, 20, 763–769.
56. Srivastava, H. M. & Gunerhan, H. (2019). Analytical and approximate solutions of fractional-order susceptible-infected-recovered epidemic model of childhood disease, *Math. Meth. Appl. Sci.*, 42(3), 935–941.
57. Atangana, A. & Owolabi, K. M. (2018). New numerical approach for fractional differential equations. *Math. Model. Nat. Phenom.*, 13(3), 1–21.

Index

A

AB fractional derivative, 85, 171–172, 234
Acquired Immunodeficiency Syndrome, 149
Adam-Bashforth-Moulton method, 156
Adams-Bashforth method, 187

B

birth parameters, 76

C

calcium signalling, 211
Caputo-Fabrizio derivative, 110
Caputo fractional derivatives, 110, 155
CD4[+] T cells, 149
cell cycle, 132
cerebral arteriovenous malformation, 21
childhood disease, 234

E

edge detector, 48–49
endoplasmic reticulum, 216
existence of solutions, 235

F

fixed-point theorem, 134, 186, 235
fractional conformable derivative, 6
fractional epidemic system, 123

H

HIV/AIDS epidemic model, 150

I

iterative perturbation method, 140

L

Laplace transform, 213
Lipschitz condition, 235
lognormal distribution, 201

M

malaria disease, 84
medical images, 2, 21
medulloblastoma, 30
meningioma, 26
Mittag-Leffler function, 244

N

numerical approximations, 187, 240
nutrition function, 76

O

oblivion function, 57

Q

q-homotopy analysis transform method, 85

R

reproduction number, 115
Riemann-Liouville integral, 213

S

SIR epidemic model, 232
SIRS-SI malaria disease model, 86

249

stability analysis, 111, 115, 189
stochastic approach, 201

T

tumour cells, 132

V

variable-, fractional-order backward difference, 58

W

WHO population growth data, 77